I0797961

David Suzuki

WITH IAN HANINGTON

LESSONS FROM A LIFETIME

90 Years of Inspiration and Activism

DAVID SUZUKI INSTITUTE

GREYSTONE BOOKS

Vancouver/Berkeley/London

26 27 28 29 30 5 4 3 2 1

Greystone Books Ltd. | greystonebooks.com
David Suzuki Institute | davidsuzukiinstitute.org

Cataloguing data available from Library and Archives Canada
ISBN 978-1-77840-360-6 (cloth)
ISBN 978-1-77840-361-3 (epub)

Editing by Ian Hanington
Copy editing by Crissy Boylan
Proofreading by Alison Strobel
Cover and text design by DSGN Dept.
Front cover photograph courtesy of UBC Archives Photograph Collection
Back cover photograph courtesy of *CBC News: The National*

Printed and bound in Canada on FSC® certified paper at Friesens. The FSC® label means that materials used for the product have been responsibly sourced.

Greystone Books thanks the Canada Council for the Arts, the British Columbia Arts Council, the Province of British Columbia through the Book Publishing Tax Credit, and the Government of Canada for supporting our publishing activities.

EU Safety Information: Easy Access System Europe, Mustamäe tee 50, 10621 Tallinn, Estonia, gpsr.requests@easproject.com.

Greystone Books gratefully acknowledges the xʷməθkʷəy̓əm (Musqueam), Sḵwx̱wú7mesh (Squamish), and səlilwətaɬ (Tsleil-Waututh) peoples on whose land our Vancouver head office is located.

CONTENTS

1 | MY HAPPY CHILDHOOD IN RACIST BRITISH COLUMBIA 1

2 | COLLEGE AND A BURGEONING CAREER 11

3 | A NEW CAREER 25

4 | STAND-UPS AND FALL-DOWNS 35

5 | FAMILY MATTERS 43

6 | HAIDA GWAII AND THE STEIN VALLEY 53

7 | ADVENTURES IN THE AMAZON 63

8 | PROTECTING PAIAKAN'S FOREST HOME 71

9 | DOWN UNDER 81

10 | STARTING THE DAVID SUZUKI FOUNDATION 95

11 | UP AND RUNNING 109

12 | RIO AND THE EARTH SUMMIT 121

13 | KYOTO, PARIS, AND CLIMATE CHANGE 131

14 | REFLECTIONS ON SCIENCE AND TECHNOLOGY 143

15 | A CULTURE OF CELEBRITY 155

16 | THOUGHTS AS I GROW OLD 163

17 | REFLECTIONS ON LESSONS LEARNED 179

1 | MY HAPPY CHILDHOOD IN RACIST BRITISH COLUMBIA

MY GRANDPARENTS ABANDONED JAPAN between 1902 and 1904, driven out by poverty to seek opportunity in Canada. After my birth, my father's parents never went back to Japan. My mother's parents were disillusioned by their treatment in Canada and returned after the Second World War to a shattered nation. They were dropped off in Hiroshima, where both were dead in less than a year.

My father and mother were born in Vancouver in 1909 and 1911, respectively, and survived the trauma of the Great Depression thanks to hard work and a strong extended family. Education was a priority for their parents, and Mom and Dad both completed high school, which was considered a good education in the 1920s.

After they married, my parents received financial help from Dad's parents to start a small laundry and dry cleaning shop in Marpole, a Vancouver neighbourhood. We lived in the back of the shop. Mom had a miscarriage early in their marriage, and Marcia and I arrived in the world on March 24, 1936. Aiko arrived a year and a half later.

During the war, when we were living in internment camps in the British Columbia Interior, Dad had been separated from the family for a year while living in a road camp building the Trans-Canada Highway. He

Dad and the twins, Marcia (left) and me, in 1936.

managed to make his way to Slocan City, where we were imprisoned, for a couple of days before going back to the road camp. Nine months later, our youngest sibling, Dawn, was born.

After the war, we were expelled from B.C. and eventually moved to London, Ontario. Marcia and Aiko took off for Toronto as soon as they finished high school to become independent. I was with Aiko when she died on December 31, 2005. My wombmate, my twin sister Marcia, was not expected to survive infancy but did and lived a long and good life and died in April 2025 at age eighty-nine. Dawn, my youngest sister, had the benefit of two sisters who had disobeyed my father's expectation that girls finish high school, find a job, and get married. Dawn graduated from university and danced with the Martha Graham Dance Company in New York for several years before getting involved with a cult.

I DIDN'T KNOW JAPAN had attacked Pearl Harbor in Hawaii on December 7, 1941. The treachery implicit in Japan's "sneak attack" against the United States Navy and the terrible war that followed threw my family and some twenty thousand other Japanese Canadians and Japanese nationals into a turbulent sequence of events, beginning with Canada's invocation of the iniquitous War Measures Act, which deprived us of all rights of citizenship.

Marcia and me on our first day of kindergarten, September 1941.

One day in early 1942, my father was gone. Left with three young children, my mother had to sort through our possessions before we made the long train ride to our destination in the Rocky Mountains. I didn't wonder why everyone on the train was Japanese. Our destination was Slocan City, where we were surrounded by hundreds of other Japanese Canadians housed in rotting buildings with glassless windows.

There was no school for the first year, and for a six-year-old kid suddenly plunked down in a valley where the rivers and lakes were filled with fish and the forests with wolves, bears, and deer, this was paradise. When school was finally opened, I began in grade one but was quickly skipped to grade two, then three, and passed to grade four at the end of the school year. I was one of the few kids whose parents were born in Canada. I never learned to speak Japanese, so I was bullied by some children. It took a long time to overcome my mistrust and resentment of Japanese Canadians because of the way I was treated in those camp days.

We were later moved from Slocan to Kaslo, a small town on Kootenay Lake less than 160 kilometres from our Slocan Valley camp. For the first time, I attended a school with lots of white kids. I shied away from them, content to explore this new area of lakes and mountains by myself.

We finally left Kaslo on a long train ride across the prairies to a Toronto suburb where Japanese Canadians were kept in a hotel until we found places to go. Dad eventually found work for himself and Mom on a hundred-acre peach farm in Essex County. We were supplied with a house,

My twin, Marcia Aoki, and me, 2024. Tara Cullis

and my sisters and I attended a one-room schoolhouse in Olinda. We were the only non-white kids in the area, but the teacher had prepared the school for our arrival and the kids welcomed us. I loved that year in Olinda. The next year, we moved to the town of Leamington when Dad found a job in a dry cleaning plant.

Leamingtonians boasted that "no coloured person stays in Leamington beyond sunset," a warning to Black people who would come from nearby Detroit to fish on the dock in Lake Erie. We were the first people of colour to move into Leamington.

Ignorance and the relentless propaganda during the war, portraying buck-toothed, slant-eyed "Japs" in the cockpit of a plane on a kamikaze mission, must have caused mystery and fear just as today's image of a Muslim extremist strapped with explosives. Every time I looked in a mirror, I saw that stereotype. I still don't like the way I look on television and don't like watching myself on my own TV programs.

Dad took this picture of me at Beatrice Lake (now part of Valhalla Provincial Park). We were forbidden to fish for some reason. Clearly, it didn't stop us.

I graduated from Mill Street Public School to enter grade nine in the only high school in Leamington. I loved the school and begged my parents to allow me to finish my first year there after they decided to move to London, about one hundred miles away. They arranged for me to stay at a farm run by friends, the Shikaze family, whose two boys were my age. We would be up at six AM to do chores; then after a hearty breakfast, we caught the school bus. After school we had more chores. It was hard work, but I never found it oppressive as we developed games. The parents spoke little English, so I learned a smattering of Japanese.

By the time I arrived in London, my parents had purchased a lot, and Dad's two brothers had pitched in and helped to build a small house. I had begun working as a framer for Suzuki Brothers Construction and loved it, working on weekends, holidays, and during the summers. By the time I arrived for grade ten at London Central Collegiate, social circles were well established and I was a stranger, a hick from a farm, an outsider. To exacerbate my isolation, I was a good student, which in that era was like having leprosy.

My loneliness during high school was intense. My one solace was a large swamp that was a ten-minute bike ride from our house. But I spent most of my waking hours daydreaming, creating a fantasy world in which I was endowed with superhuman athletic and intellectual powers that would enable me to bring peace to the world and win mobs of gorgeous women begging to be my girl.

A carp caught in the Thames River in London, Ontario.

In his bestseller *Is There Life After High School?* (1976), Ralph Keyes divides high school populations into the "Innies" (sports players, cheerleaders) who set the social standards while the "Outies" (glee club, debate society, band, etc.) mostly wish they were Innies. In my last year of high school, one of my fellow nerds suggested I run for school president. I said no. When I told my father, he was disappointed and asked why. "Because I'd lose," I explained. Dad was outraged. "How will you know if you don't even try? There is no shame in trying and losing. The shame is not even trying," he said.

So I went back to my friend and said I'd give it a try. My public speaking experience in oratorical contests served me well during the campaign at Central, and I appealed to the Outies. To my amazement, I won with more votes than all the other candidates combined. It was another powerful lesson: There are a lot more Outies than Innies, and together that means power.

For me, the alienation that began with our evacuation from the British Columbian coast and continued through high school has remained a fundamental part of who I am, despite the acquired veneer of adult maturity.

Dad's Model A decked out for my campaign to be student president in 1953. (Note the box on the back where we carried leaves to use for compost.)

George Monbiot

AUTHOR, JOURNALIST,
GUARDIAN COLUMNIST

Throughout my adult life, David has been a guiding light for me, an inspiration and an example of the three most important traits in anyone who wants a better world: courage, persistence, and resilience. While in all walks of public life, we are confronted with constant pressure to conform, to censor ourselves, to accede to the demands of power, David has resisted throughout his ninety years, confronting the forces that are destroying our life-support systems and showing us how we might live in harmony with Earth, our Mother. Scarcely anyone on the planet has made such a contribution. I honour him.

Faye Simon

HIGH SCHOOL
CLASSMATE

In 1949–50, the student body of my high school in Leamington, Ontario, was divided into four houses: Alpha, Beta, Delta, and Gamma. Each house planned activities and events outside the academic curriculum, such as hosting an annual dance and competitions in public speaking, attendance, punctuality, and others.

I met David Suzuki in 1949 when I was Beta vice-captain. He contributed in many ways, and his views on nature were clear even then. In later years, David spoke about the importance of protecting Earth and people by changing products and practices then in use. There was pushback from farmers who felt their livelihoods were being threatened, but David persisted. Over time, governments began to legislate for more control over things like pesticides, and the agricultural industry began to shift in favour of environmental protection.

I once saw David making a guest appearance on a Canadian TV comedy show. I nearly fell off my chair laughing when the man I remembered as a "reserved boy" and later a serious activist showed up wearing a strategically placed fig leaf. When he was asked about it later, his response showed his self-denigrating sense of humour when he said something about "this old hulk."

I am thankful that this remarkable man has been a positive influence throughout my life.

PUBLIC SPEAKING WINNERS

Back Row: Maurice Cosyn, David Suzuki, Loanne Graham, Douglas Graham, George Cole. Front Row: Jane Anderson, Joyce McGregor, Faye Hillier.

PUBLIC SPEAKING

Again this winter term Public Speaking played an important part in our school activities. Evidence of enthusiasm was proven by the fact that 163 students entered the eliminations from the four houses.

The students are quick to realize the advantages of the art of speaking and the fine training received at the High School.

In the final contests, March 6 and 7, two speakers from each of the houses —Alpha, Beta, Gamma and Delta—vied for the Public Speaking Crest. As this crest is considered one of the most important, competition is very keen.

The crest this year was won by Alpha House. The speakers who aided in winning the crest for their house were:—Maurice Cosyn, Jane Anderson, Joy Morrison.

Every year since its inception in 1939, enthusiasm over Public Speaking has increased and the final contests have become increasingly closer. This year the students witnessed the closest contest ever judged. The margin of points between the houses was very small, Alpha winning by a scant half point.

Thus this year we salute Alpha house for being the proud possessor of the Public Speaking Crest for 1949-1950.

Faye Hiller (now Simon) and I won public speaking awards in high school.

2 | COLLEGE AND A BURGEONING CAREER

THANKS TO A CHANCE ENCOUNTER with John Thompson, a fellow classmate at Central whose father was the dean of business at the University of Western Ontario and who went to Amherst College in Massachusetts after grade twelve, I ended up also attending college at Amherst. "It's a great school," he enthused. "You ought to go there." He arranged to have the application forms sent to me. I was accepted with a scholarship of $1,500, which at that time was more than my father earned in a year. I became the first person in my family to attend and graduate from university. My parents had pounded home the importance of education as a means for us to escape the extreme poverty we found ourselves in after the war.

In grade twelve, I had asked the prettiest Japanese girl in London, Joane Sunahara, to go to a New Year's Eve dance. She was a great dancer and a terrific kisser. The next year when I became student president at Central Collegiate, she was elected student vice-president at Tech, so we were a couple at all the social events at the two schools. But once we had graduated, she went to Ryerson (now Toronto Metropolitan University) in Toronto to train as a lab technician, and we understood that we would stay in touch but also date others.

Harvesting salmon pituitary glands for future Nobel Prize winner Michael Smith's biochemical studies. (Yup, I used to smoke.)

In fall 1957, my senior year at Amherst, an Asian flu epidemic swept the world. Like many, I succumbed to the virus and staggered to the school infirmary. I collapsed into bed, with only the radio as a diversion. I was jolted out of my stupor when an announcer interrupted the programming to inform us that the Soviet Union had successfully launched a basketball-sized satellite called Sputnik into space. I had had no inkling that there was a space program, and the feat captured my imagination. But in the months that followed, I and the rest of America agonized as the United States initially failed in spectacular fashion to get a satellite into orbit. Meanwhile, the USSR announced one first after another: Laika, the first animal (a dog) in space; Yuri Gagarin, the first man; the first team of cosmonauts; the first spacewalk; Valentina Tereshkova, the first woman.

The U.S. was determined to catch up and poured money into student support, university science departments, and government labs. Even though I was Canadian, I later received funding to carry on with my graduate studies at the University of Chicago.

My professor Bill Baker, fellow PhD student Anita Hessler, and me in the fly lab at the University of Chicago.

I had always wanted to be a biologist. In my early years, I dreamed of being an ichthyologist, someone who studies fish. Later, when I became an avid collector of insects, I considered entomology as a possible profession. In the 1950s, most people who did well in sciences went into medicine. At the end of my sophomore year at Amherst, I took my transcripts to the dean of medicine at the University of Western Ontario, who assured me I would have no problem getting into med school. My future profession seemed set.

But in my third year of college, as an honours student in biology, I was required to take a course in genetics and fell madly in love with the elegance and mathematical precision of the discipline. In my senior year, I decided to switch from a career path as a doctor to genetics. It took years for my mother to accept that. By the time I decided to go to grad school in the fall of my senior year, it was too late to apply for fellowships. Curt Stern, one of the world's top geneticists, accepted me as his student at the University of California at Berkeley. But Joane and I planned to get married, and I couldn't afford to go without financial support. Bill Hexter, my thesis adviser at Amherst, called a friend, Bill Baker, a fruit fly geneticist at the University of Chicago, who offered me a position as his research assistant supported by his grant.

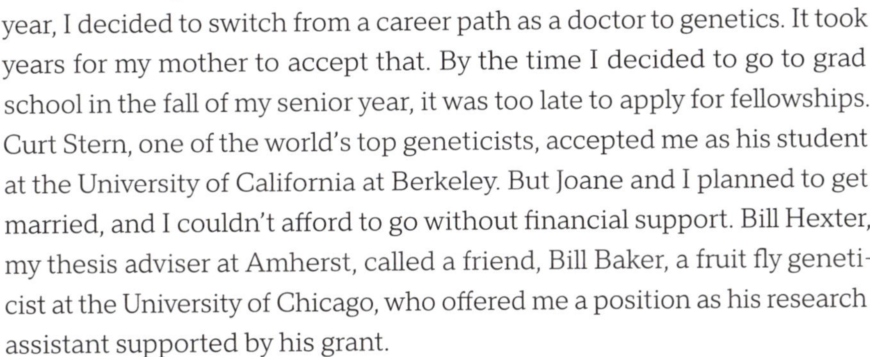

When I graduated from Amherst in 1958 with an honours degree (cum laude) in biology, I knew I could at least be a good teacher, but in graduate school at the University of Chicago, I felt a burning desire to do experimental science. I enrolled as a student in the zoology department, while Joane, whom I had married in August 1958, worked as a technician. Tamiko was born in January 1960. I had not grown up with babies in my life and discovered the incredible joy of fatherhood, but it added to our

I married Setsuko Joane Sunahara in 1958.

financial needs so I completed my doctorate in zoology in less than three years.

In June 1961, I received my PhD. My thesis adviser, Bill Baker, had worked for years in the biology division of the Oak Ridge National Laboratory (ORNL) in Tennessee and strongly recommended that I apply for a position there. I did, and I was delighted to receive my first full-time job as a research associate. It was in the lab of Dan Lindsley, one of the world's experts in chromosome mechanics, an arcane field of fruit fly genetics involving the manipulation of chromosomes.

Oak Ridge was established in 1943 as part of the Manhattan Project to develop the atomic bomb. Ironically, the institution that had supplied material for the bombs that had demolished Hiroshima and Nagasaki was now a hotbed of world-class research and international cooperation, and I was part of it. Tennessee had been a pro-slavery state and a part of the southern Confederacy in the Civil War. Overt signs of racism remained. Because of my own experience during the war, I identified strongly with the Black community. In Dan Lindsley's lab, the chief technician was Ruby Wilkerson, a Black American who lived with her husband, Floyd, in the nearby village of Philadelphia. I started attending meetings of the local chapter of the National Association for the Advancement of Colored People (NAACP). I deeply resented the way Ruby was treated in Oak Ridge and soon was blaming all white people. "David," Joane said one day, "you're becoming a racist. All white people aren't like that."

Although I could have stayed on at Oak Ridge as a research associate and had been offered several positions in California, I felt deeply

My Oak Ridge lab companion, Ruby Wilkerson (centre); her husband, Floyd; and her daughter, Patricia, with trout at Dad's pond near London.

estranged from the culture because of the racism. Canada and the U.S. had moved people of Japanese origin away from the coast and incarcerated most in hastily set up camps, but Canada was smaller, and I believed there was more of a chance to work for a better society. Canada was different, not better, than the U.S. The opportunities for a scientist in the U.S. were much greater at that time, but I have never regretted my decision to return home.

A position as assistant professor arose in the genetics department at the University of Alberta. Edmonton was an excellent place to begin my career. When I arrived in the summer of 1962, I would work at the lab until two or three in the morning and was thrilled that it stayed light so

Hamming it up in the lab at UBC. CBC Still Photo Collection, CBC Licensing

late because of Edmonton's northerly latitude. However, that winter, the thermometer plunged to -40°C, a temperature I had never experienced and did not wish to again. When a position came up at the University of British Columbia (UBC), I applied and was invited for an interview. I took the job.

When I took the position in Edmonton in 1962, I had applied for a research grant from the National Research Council in Ottawa and was shocked to be awarded only $4,200. I was told that a beginning assistant professor normally started with a $3,500 grant; I had been given a larger amount because of my postdoctoral year at Oak Ridge. My American peers were getting grants of $60,000 to $80,000. Canada had not moved into the post-Sputnik era, as the U.S. had, with a huge commitment to science as part of the Cold War competition. I wondered whether I could remain in Canada and still succeed as a scientist.

As I was leaving Oak Ridge in 1962, George Stapleton, an administrator I had come to know, advised me to apply to the U.S. Atomic Energy Commission for research money. Once I got to the University of Alberta, I did. To my surprise and delight, I received a substantial grant. It is such an irony that the U.S. gave me, a foreigner, the support that enabled me to remain in my own country.

I was in my late twenties when I arrived at UBC in 1963. Not surprisingly, as I spent more and more time in the lab, Joane and I had less and

Young professor at UBC in 1968, able to speak out because I had tenure.

less time together. Besides Tamiko, we now had Troy, born in 1962. Joane had worked hard, even with a child, so that I could go to graduate school, and now that I was settled as a faculty member, we should have had more opportunities to be together. But I was too ambitious to give up the time. Soon after the birth of Laura, our third child, in 1964, Joane and the children moved into a home we had just bought. I did not.

When I returned to Canada from the U.S., I was consumed by a passion to probe chromosome behaviour during cell division in *Drosophila*. I also enjoyed teaching and put a lot of time and energy into it. Student questions forced me to read up on the history of genetics, which I hadn't been taught in college. It was devastating to discover that geneticists early in the twentieth century had extrapolated from their studies of the heredity of physical characteristics in mice, fruit flies, and plants to make pronouncements about the heredity of intelligence and behaviour in humans. I began to speak out about the potential abuse of genetics. This did not sit well with my colleagues in that field.

An outsider sees things from a different angle and thus often recognizes what others may not see. A scientist working in biotechnology with the prospect of making a lot of money from a product can be resistant, if not blind, to questions of risks that someone without a vested interest might see with greater clarity. For me, status as an outsider has been a mixed blessing. When I was younger, I so wanted to fit in and not stand apart, to be accepted and liked. But by being on the outside, not only do I see things from a different perspective, I also don't have a stake in the status quo or in companies, groups, or organizations of which I might be critical.

It is easy to rationalize any side of an argument. When the powerful tool of recombinant DNA that enabled gene transfers from one species to another began to be exploited extensively, I banned the technique from my lab so that I would be free to enter discussions about its benefits and hazards without the conflict of wanting to protect my own work.

Tamiko Suzuki

ELDEST DAUGHTER

I am the oldest child of David's first family, so I have memories of growing up in the '60s and '70s when he was building his career as a professor in genetics at the University of British Columbia. I remember playing in his lab and running up and down the halls and thinking none of my friends' dads had such a cool workplace. Dad would show me the *Drosophila* he was working on (he always used the scientific name) under the microscope—where flies would fall down, paralyzed at certain temperatures. I was suitably impressed without understanding the significance of what I was seeing. I just admired seeing the flies up close with their beautiful wings and eyes. I was weirded out by the strains with legs growing out of their heads, but mostly I was just flattered to be invited into his world of grown-up stuff. I would then break the mood by reminding him to bring some of his extra *Drosophila* home to feed to my pet spider—the one with the beautiful web who lived out on the porch.

When he would come home, he would tell us to gather our buckets and nets and get on our gumboots, and we'd walk to the local swamp two blocks away. I remember friends and teachers warning us not to go to the swamp because "the boogeyman lived there," "there's quicksand," and "children disappear and are never seen again." I knew this couldn't be true because I would spend so many wonderful hours there with my dad and brother, collecting frogs' eggs, water beetles, and fairy shrimp, picking bulrushes, and chasing butterflies and dragonflies. It was a magical place that, sadly, was drained in the '70s and made into a track and field. I laugh when

I walk by there now in winter because the grass is always a boggy mess. The land remembers what it used to be for thousands of years.

Those times in the swamp taught me the magic of being in nature. Nature was never scary or dirty or boring, and that lesson is something I tried to pass on to my kids and their friends. Dad's trips to the swamp also taught me not to 100 percent believe whatever teachers or parents said—they either had an agenda or they just didn't know better!

Dad took us camping and hiking a lot in those days. We used heavy canvas tents that leaked and old army sleeping bags. It was always fun but was often cold and uncomfortable at night. As an adult, I spent almost every weekend backpacking in the summer and alpine touring in the winter, so I was appreciative of nylon tents, Gore-Tex jackets, and warm sleeping bags!

Dad usually brought some of his students along to share the fun (and perhaps help with the babysitting). I was familiar with the smell of pot as a kid, which demystified it for me and meant "smoking up" never represented "rebellion" or "coolness." To my little kid eyes, people just became boring and silly when stoned so I wasn't eager to try it in high school or university. I do use it medicinally now to deal with my arthritis and to get to sleep, so maybe the apple hasn't fallen far from the tree.

Laura, Tamiko, and Troy, 1966.

John Bergeron

EMERITUS ROBERT REFORD PROFESSOR AND PROFESSOR OF MEDICINE AT MCGILL UNIVERSITY

Kathleen Dickson

RETIRED CHIEF TECHNICIAN AT THE MONTREAL NEUROLOGICAL INSTITUTE

When the University of Alberta announced it would award David Suzuki an honorary doctor of science degree at its 2018 spring convocation, there was a backlash.

Critics argued his stance against Alberta's oil sands made him an unsuitable recipient. Although David was receiving the honour because of his efforts to boost science literacy and environmental awareness, it's important to remember that before he became a broadcaster and activist, he was a globally recognized scientist. David embarked on a career in science with his first professorial appointment in the genetics department at the University of Alberta in 1962.

As a dedicated scientist and communicator of science, David frequently presents the scientific case for climate change and the effect of fossil fuels to the public. No other scientist in Canada has been pilloried so strongly for simply presenting the data. While many may think of David as an activist or a television presenter, he also holds a legacy in genetics research.

As a professor at the University of British Columbia, David wanted to understand how muscle worked and tried to uncover genes that would cause paralysis. He reasoned that such genes would be in common to all life forms with muscle. He selected the common fruit fly, *Drosophila melanogaster*, for his experiments.

In 2018, I was awarded an honorary degree from the University of Alberta, where my genetics and broadcasting careers started. It was a controversial move in the centre of oil country!

This was genius!

The fruit fly *Drosophila melanogaster* is a model organism used to understand the biology of other organisms, including humans. David's experimental design was elegant and conclusive. He simply fed fruit flies a chemical known to cause random mutations.

In 1967, David was the first to use temperature to screen for genetic mutations in fruit flies. These temperature-sensitive mutants behaved normally when the flies were kept near room temperature (22°C). The effects of the mutation were only observed at a higher temperature (29°C), he later wrote in *Science* magazine.

At the high temperature, some of the flies became paralyzed and fell to the bottom of their container, while unaffected ones simply flew. When he changed the temperature to 22°C, a small number of the paralyzed fruit flies at the bottom of the container regained the ability to fly.

The flies' recovery after the temperature change meant they harboured a mutation in a single gene. David named the gene *shibire*, the Japanese word for "paralyzed."

It was not long before the international community of discovery researchers followed David's lead. He had correctly predicted that studying genetic mutations in fruit flies could help scientists identify genes involved in the development of human disease and other phenomena.

History will judge the outcome of David's attempt to use observation and reason to, as the University of Alberta explained, "spread scientific literacy, appreciation of nature, and knowledge of the deep ecological crises threatening life on the planet." But it remains that David may be considered Canada's pioneer in fundamental genetics research.

Thomas Grigliatti

FORMER COLLEAGUE, PROFESSOR OF CELL AND DEVELOPMENTAL BIOLOGY, UNIVERSITY OF BRITISH COLUMBIA

David is an excellent scientist. Additionally, and even more importantly, he is an outstanding communicator. He can explain both simple and complex concepts clearly and in terms that are understandable to most of us, regardless of age. He devoted himself to using the combination of these two exceptional talents to inform and educate us all about the fragility of the environment and the consequential risks and dangers of escalating changes in our natural environment and the resultant rapid climate changes.

That was a calling, a monumental undertaking, and a contribution to society that far exceeds the contribution of the hundreds of scientific investigators toiling away in their respective laboratories, whether they work as academics, for government agencies, or commercial enterprises. His tireless work has educated millions of people about these changes and their likely consequences. In addition, it has spurred a lot of research in climate change, and it has motivated many others to join in with this education and call for change.

Lastly, but just as importantly, while we celebrate David's ninetieth birthday and retirement from the CBC, we also know that his accomplishments in the establishment, growth, management, and success of the David Suzuki Foundation are equally due to his partnership with Tara Cullis.

3 | A NEW CAREER

IN 1954, WHEN I GRADUATED from high school and went away to college, my family had never owned a television set. At that time in London, Ontario, television was still a novelty, and a pioneer who purchased a TV set required a giant antenna to pick up signals from Cleveland or Detroit. On another front that would turn out to be a thread in the fabric of my life, my father had encouraged me to take up public speaking while I was in high school.

Amherst College in the 1950s aimed to graduate students who were well rounded in the humanities, sciences, and physical education, and men had to take public speaking in their sophomore year. I took it seriously and won top marks in the six speeches we had to give over two semesters. As well, as an honours biology student, I was required to make a scientific presentation to students and professors in each semester of my senior year, and I discovered I had an ability to present complex scientific topics in a way that not only was understandable but also excited the listeners.

After I arrived at the University of Alberta in 1962 to take up my first academic position, word got out that I was a good lecturer, and I was invited to give a talk on a program called *Your University Speaks*, broadcast on a local television channel. It featured university professors lecturing

Being interviewed by students in the late 1960s.

on subjects in their areas of expertise, aided by slides. I accepted the invitation and did so well that I ended up appearing on eight episodes. The series was broadcast early Sunday mornings, so I was shocked when people stopped me and told me how much they had enjoyed one of the shows I had done.

I moved to the University of British Columbia a year after returning to Canada. In Vancouver, I was asked to appear on television to do the occasional book review or commentary on a scientific story. I became more interested in the medium as a way to communicate. I proposed a television series about cutting-edge science to Knowlton Nash, head of programming at the Canadian Broadcasting Corporation (CBC), and he approved the series to come out of Vancouver. The show, *Suzuki on Science*, was broadcast across the country in 1969 and was my first involvement in a television series with a national audience. It ran for two seasons and was renewed for a third, but I quit. We had a lousy time slot and low budget, and I saw no future for the show.

In 1974, Jim Murray, long-time executive producer of *The Nature of Things* on CBC Television, began a new show called *Science Magazine*, a weekly half-hour collection of reports on science, technology, and medicine. He had known about my Vancouver-based series, and after we met and talked, he hired me to be the host of the new one. It was an immediate hit, drawing an audience 50 percent larger than the long-running popular series *The Nature of Things*, which had begun in 1960. I took a leave from UBC to host the program.

Halfway through that first season, Knowlton Nash informed us the series would be dropped. On the last episode, I thanked viewers for watching and told them the show was ending. Viewers responded with

Television host of *Suzuki on Science*, photographed in 1969.

letters and phone calls objecting. The series was restored and ran for four more successful years.

During the first season of *Science Magazine*, Diana Filer, executive producer of the CBC Radio series *Concern*, attended a speech I gave at the University of Toronto. She had proposed a new radio science series called *Quirks & Quarks* and hired me to host it when it went to air in 1975. I was host of both *Science Magazine* on television and *Quirks & Quarks* on radio, which was a full-time job, until 1979.

Jim Murray was generous in showing me how to make high-quality films, and we became good friends. In 1979, he proposed me as the host of *The Nature of Things*, which had been reformatted and was to become an hour-long program called *The Nature of Things With David Suzuki*. I couldn't do both that and *Quirks & Quarks*, so I left radio with great reluctance. It was the medium I enjoyed most, because interviews were relaxed, and there was an opportunity to be spontaneous, humorous, and even risqué, since tapes could be edited and still retain warmth and intimacy. In contrast, television is controlled, because airtime is so valuable.

When I started out in television in 1962, I never dreamed that it would ultimately occupy most of my life and make me a celebrity in Canada. I thought I might have a knack for translating the arcane jargon of science into the vernacular of the lay public, and I felt that speaking on television was a responsibility I had assumed by accepting government research grants and public support at a university. When I appeared on a television show for the first time, I enjoyed the novelty; I did not anticipate the notoriety that would come from being onscreen regularly.

I realized how important the applications of scientific ideas and techniques were to people's lives, and I thought my role was to make those applications accessible to the public. I wanted to empower the public, but the opposite happened because of the nature of the medium. Regular viewers of *The Nature of Things With David Suzuki* watched the program on faith that what we presented was important and true, and they came to expect me to tell them what to do or to act on their behalf. If I phoned a politician's office, even the prime minister's, chances were very good that my call would be returned within half an hour—not because I'm an

important person, but because an informed politician knows that a million and a half people watched my shows regularly.

As the head of a large research lab, I was constantly at the centre of activity. If not actually carrying out an experiment myself, I would be having discussions with various members of the team, reading new publications, arguing about what we should be doing next, talking about student projects, and so on. What a contrast with making a television program. Although a shoot involves moments of intense activity and concentration, those are punctuated by long periods of sitting around waiting, and the host is the least important factor.

I got one important lesson during the early days of *Science Magazine*. We were in a lab in New Orleans. I was seated on a stool while the crew rushed about. The cameraman, Rudi Kovanic, came in and said, "Oh no. This doesn't look sciencey at all." He started moving equipment around and filling flasks with different coloured water. Then the lighting man came in to fiddle with light on the background, then on me. The soundman pinned a microphone up my shirt, then turned on his machine to listen to my voice counting from one to ten and to hear whether my clothes were causing any scratching on the mike. Finally, all was ready. Rudi came up and put a light metre against my nose to take a reading, then scurried back to the camera and adjusted

Publicity shot for *The Nature of Things With David Suzuki*. CBC

the focus. Then he came back and did the same thing again. Nothing had changed, but Rudi took a third reading and then a fourth. I finally said, "For Christ's sake, Rudi, shoot the damn thing!"

Jim came over, grabbed me by the shoulder, shoved me into another room, and shut the door. "Listen, Suzuki," he hissed. "Those guys are busting their asses to make you look good. And that's not easy." I realized he was right. I get all the credit for what they're doing. No viewer thinks, "Gee, the lighting was perfect" or "The sound was incredible." They just think, "Suzuki's show was so interesting." No longer the hotshot scientist, a chastened on-camera person slunk back to his stool and never complained again about the drawn-out challenge of making him look good!

At the CBC Atrium in Toronto at my retirement send-off from *The Nature of Things*, May 2023.

Ivan Fecan

MEDIA EXECUTIVE PRODUCER, PHILANTHROPIST, FORMER PRESIDENT AND CEO OF CTVGLOBEMEDIA, FORMER CEO OF CTV

In 1975, David was already a celebrated geneticist and media figure; I was a twenty-one-year-old university dropout who had no business being his producer.

In early spring of that year, one of CBC Radio's legendary producers, Diana Filer, convinced David to host a weekly national science program that she would then pitch to management. The idea they developed was to go deep and explore the important ethical and moral choices facing science and how individual scientists were approaching those thorny issues. Given David's concerns about the possible misuse of genetics, he would be the perfect host and interviewer.

I was freelancing at CBC Radio. Diana got the green light and hired me to work on the show. We started pre-production—this is when you plan the season in terms of what topics to cover, what interviews to book, etc. And then Diana got lured away by television. A few months away from our airdate, we had David, we had a unique editorial concept with some roughed-in ideas, we had a catchy title—but we had no Diana!

David eyed me warily. He decided to give "us" a shot. A battlefield promotion, as it were. And we made it work. In that first season, David did some amazing interviews. I particularly recall the one with Edward Teller, known as father of the atomic bomb. We leavened the serious stuff with gee-whiz moments; for instance, Isaac Asimov did a weekly column.

We launched in fall 1975, and it was a hit. We exceeded everyone's audience expectations, got good reviews, and won awards. *Quirks & Quarks* is still on the air. David taught me so much, particularly about leadership. I am forever grateful to him for taking a chance on me.

Ivan Fecan and I launched the CBC Radio program *Quirks & Quarks* in 1975. CBC

Sue Dando

FORMER EXECUTIVE PRODUCER, NOW CONSULTANT TO CBC'S *THE NATURE OF THINGS*

When I became executive producer of *The Nature of Things* in 2012, I didn't know what to expect from my first meeting with legendary host David Suzuki. My time with the star of the series was scheduled for a tight forty-five minutes while he was briefly in Toronto. We ended up chatting for more than two hours—about the show, of course, but also about our families and books and current events and food. David asked me as many questions about my life and views as I did about his.

That wide-ranging and engrossing conversation was one of many as we worked together for the next thirteen years. I soon realized that David saw everyone he worked with as a partner and colleague engaged in a great enterprise: creating science stories to help viewers better understand our world.

Since everyone who worked with David knew there was nothing our host wouldn't do for the cause, they concocted some imaginative scenarios during David's four decades with the series. Pose naked with a fig leaf? Yes. Take a mud bath on camera? Why not? Explore personal aspects of his own life—and his own aging body? Absolutely. There was always a point to these on-camera adventures: to make the underlying science clear and accessible. David is amazing at that.

Many people use the word *inspiring* to describe David, and I'm going to use it too. He truly is inspiring. In big ways, such as by standing up for what he believes. And in smaller ways that set a terrific

example, such as by staying fit and active, spending time outdoors each day, and nurturing a sense of curiosity and wonder.

David also has a capacity for delight that is a pleasure to witness. One thing I discovered in our very first conversation is how much I love it when David throws his head back in a huge laugh. I hope he knows how many people, including me, are cheering for him and sending him their love.

Sue Dando was the executive producer of *The Nature of Things* for many years. CBC/George Pimentel Photography

Radio Shack
TRS-80

4 | STAND-UPS AND FALL-DOWNS

IT'S ONE THING to memorize lines and deliver them before a camera; it's quite another to move or even gesture while also speaking. Add factors beyond those and the task becomes even more challenging. I am filled with admiration for David Attenborough, the British host of countless natural history television programs. His stand-ups set a high standard. Actually, it was often a sit-down rather than stand-up.

Sometimes I have to juggle several stand-ups in a shoot. I had remarried in 1972; my wife, Tara, was pregnant with our second child when, in 1983, filming started for *A Planet for the Taking*, the biggest television series I had ever been involved with. We had slotted in a three-week interval around the time the baby was due when I could be in Vancouver. The anticipated date for the baby's arrival came and went, and day after day the amount of time I had available to stay home shrank.

We had three camera crews out filming at the same time—one in India, the other two in Europe—and I was absolutely needed to do the stand-ups because they would hold the entire series together. If I couldn't be there when filming was going on, I would have to be sent out with a crew later just to shoot stand-ups, and that would be expensive. The day I was supposed to leave for India came and still no baby. Sarika arrived three

Getting started with a Radio Shack computer. CBC Still Photo Collection, CBC Licensing

days after that, so I stayed around for another two days and then flew to India, five days late.

I did my stand-ups in India over several days, then moved on to Europe, Egypt, and Israel before flying to Kenya, where producer Nancy Archibald was filming a sequence on baboons. At this point, I had not seen Tara, Severn (our first child), or the new baby for over three weeks. Tara had received clearance from doctors to fly with Sarika (and three-year-old Severn) to meet me in England, where I would be shooting a segment on the mathematician Isaac Newton, so I had to leave Nairobi on a certain date.

Shirley Strum, an expert we hired, assured us the baboons would wake, forage for a couple of hours, and then settle down in mid-morning for long enough that we could get all my stand-ups. We found the troop as

I once appeared on the classic TV show *The Beachcombers* with my good friend Bruno Gerussi. CBC

they bedded for the night. When the sun came up, we followed them, but every time we thought they were settling down and we set up to film, they would move. We spent the entire day and only got one of the four needed stand-ups. I could see the plane leaving without me. It was crushing, and the expert expressed surprise at the baboons' behaviour. The next morning, we started again. This time, the animals behaved as predicted, we got all the stand-ups in the morning, and I was on the plane to England.

One shoot I did had an amazing effect. Actually, it was a shoot for a still photo, not for a program. In the 1970s, when an episode of *The Nature of Things* failed to get more than a million and a half viewers, we would worry. But with cable and dozens of competing channels, our numbers fell steadily until our average, while still robust for a CBC program, sank below a million. I kept saying, half jokingly, that we could get dynamite numbers if we did a program on the penis, a perfectly good subject for a science show. Jim Murray laughed as he thought I was kidding. When Michael Allder became the executive producer, I mentioned the idea and he immediately expressed interest. He commissioned the program, and it focused on the male obsession with size and some of the techniques used to enlarge the organ. The show was called "Phallacies."

Michael wanted a series of photos of me for publicity and arranged for a shoot at his cottage in Georgian Bay. As we were leaving the CBC for the shoot, Helicia Glucksman, our publicist, handed me a couple of fig leaves and said, "If you have time, please take a photo wearing this for 'Phallacies.'" It was all said lightheartedly, and I didn't know whether she was serious.

To get the best light, we shot early the next morning with the rising sun. The photographer was efficient, and we soon had all the pictures Michael wanted, so as a lark, I taped the fig leaf to my crotch and we set up a bench for me to stand on and pose. It was quite cold out so I had to drape a blanket over my shoulders between shots so I wasn't covered in goosebumps. It was a very large fig leaf, so I felt it was modest.

Helicia arranged for the photo to be on the cover of the *Toronto Star*'s *Starweek* TV guide, and I was astonished to see the reaction when it came out. It was picked up by dozens of newspapers across the country and written up as if it was incredible for me to pose that way. I did receive a couple of letters and one nasty phone call (all from women) expressing disgust at my "obscene photo." Overwhelmingly, the response seemed to be surprise that a sixty-four-year-old man could still be in reasonable shape. There was even a suggestion that my head had been superimposed on someone else's body.

I am not a bodybuilder, and at my age, testosterone levels are too low to allow me to build up muscle mass, but I had been exercising regularly for decades, ever since I married a much younger person. Once when my daughter admired a photo of someone's "abs" by saying "Wow, look at this six-pack," I had interjected, "What about mine?" Sarika retorted, "Dad, you've got a one-pack!" I have been gratified that even as I turn ninety, my body has responded to exercise. After Sarika's jibe, I developed a series of exercises for my belly, and they worked.

We got the best rating for "Phallacies" that we had had for many years, but it was bittersweet for me. Staff at my foundation had worked hard for years for every story we got into the media on environmental issues. Then I take off my clothes for one shot and we get gangbusters exposure.

FACING: **The notorious fig leaf shot for *The Nature of Things With David Suzuki* episode "Phallacies." My Haida friends ask why the leaf is so small.** CBC/Greg Pacek
ABOVE: **A publicity shot with illusionist David Ben for a segment on Martin Gardner called "Mathemagician" for *The Nature of Things With David Suzuki*.** CBC

Ziya Tong

SCIENCE BROADCASTER,
AUTHOR, FILMMAKER

When I grow up, I want to be a nonagenarian badass like my friend, the legendary David Suzuki.

David has spent a lifetime fearlessly guided by truth—scientific truth, moral truth, and the unflinching courage to speak it to power. For decades, he has stood on the front lines of environmental and social justice, challenging the status quo with intellect, heart, and fire.

It has been one of the greatest honours of my life to learn from him, to be inspired by his indefatigable spirit, and to call this extraordinary human my mentor and friend.

My daughter Severn and me enjoying time with TV host and producer Ziya Tong (centre).

Jane Fonda

ACTOR, ACTIVIST

I met David Suzuki many decades ago in the late '80s when he spoke on Earth Day at the top of a mountain above Malibu. I remember being struck by his handsome face, the lively twinkle in his eyes, and his strong body. (He was wearing shorts.) When he spoke to the large crowd assembled in a tent on the mountaintop, I was transported to a visceral understanding of the exquisite interconnectedness of the natural world. It was truly an epiphany and remains a vivid memory.

Among other things, David spoke of how the flapping of butterfly wings in one part of the world could cause winds in another part. It was the first time I had heard that theory. Other breathtaking stories he revealed then and in subsequent speeches I've attended—the salmon's relationship to forests, for example—have helped me understand something else truly essential: We humans are part of nature. Without nature there is no "us." And by treating nature as a resource for our profit rather than as a life source, we put "us" in peril. Dr. Suzuki is a poetic, soulful scientist who has changed many lives and opened many minds, mine included.

Jane Fonda gave a generous gift to the foundation at my eightieth birthday. Brendon Purdy

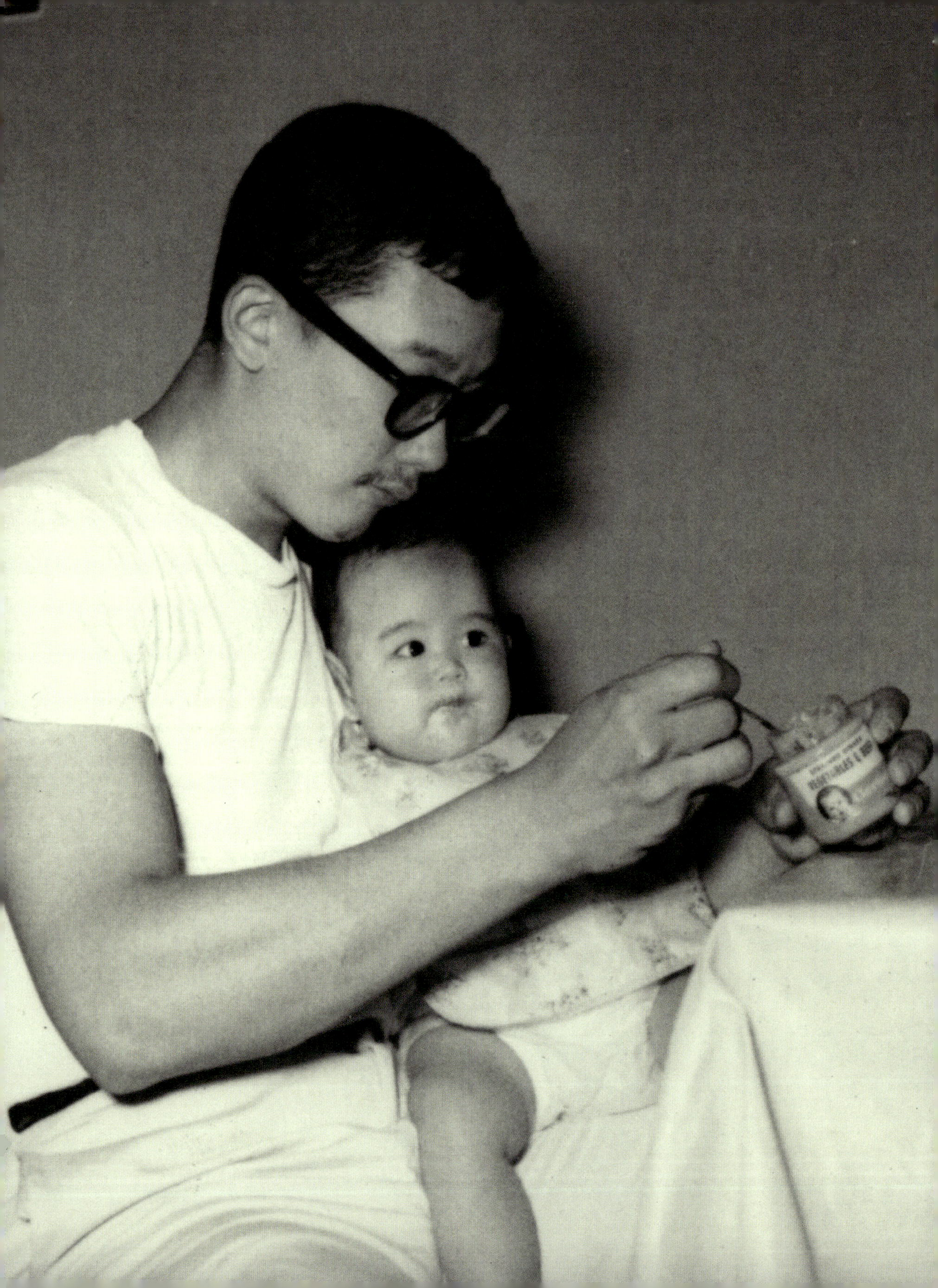

5 | FAMILY MATTERS

I WAS BECOMING MORE INVOLVED in television when Joane and I separated in 1964. By that time, we had two children and a third was on the way; we didn't divorce until two years later. Laura was conceived before Joane and I had agreed to separate. She was born prematurely, on July 4, 1964.

I moved into an apartment near the family so I could still see the kids every day. But when Michael Lerner, an eminent population geneticist at the University of California at Berkeley, invited me to teach a course there, I eagerly accepted. During my stay, the battle over People's Park broke out, and I took part in the demonstrations that ended in tear-gassing, buckshot, and death at the hands of the California National Guard, called in by Governor Ronald Reagan. I was appalled at the violent attempt by Americans to put down their own youth and realized then that my decision to return to Canada in 1962 was still the right one.

I had gone to Berkeley looking like a square, and I came back decked out in granny glasses, a moth-eaten moustache and beard, and bell-bottoms. But the University of British Columbia, like Berkeley, was swept up in revolutionary fervour and the sexual revolution. Physical appearance didn't seem to matter anymore, and I no longer felt such intense

My children are my pride and joy. With Tamiko in 1960.

self-loathing because of my small eyes and Asian appearance. In the pre-AIDS period before the 1980s, there was rampant experimentation with drugs and sex, and although I was too unhip and insecure to ever try LSD, it was widely believed I was "into" psychedelics, and I heard rumours (false) that "acid" was being synthesized in my lab.

On December 10, 1971, I was scheduled to give a talk at Carleton University in Ottawa. I entered the lecture room at the top of Dunton Tower to find it packed with several hundred students. As I began to speak, I noticed a sensationally beautiful woman who looked like Rita Hayworth sitting near the front. After I gave my speech and people began filing out of the room, a handful came to the front to carry on a dialogue. The beautiful woman was one of them. I had never acquired the self-confidence to "pick up" someone or even start a conversation in that direction. Instead, as I was leaving, I announced in a loud voice, "I hope you're all coming to the party tonight," and I left.

I had to sit on a panel early that evening and did not see the beautiful woman in the audience, so I figured I had failed. Afterwards I was driven to the party, which was packed with students, a number of whom immediately surrounded me to engage in serious conversation. About half an hour later, the woman arrived. I ducked out of the ring of people, popped up in front of her, and asked her if she wanted to dance.

The sensational woman was Tara Cullis, who was working on a master's degree in comparative literature at Carleton. She was twenty-two; I was thirty-five. I learned later she used to watch *Suzuki on Science* with her boyfriend and had attended my talk out of homesickness for British Columbia. After hearing my lecture, she felt for the first time in her life that she could imagine marrying someone—me.

How helpful that she was from B.C. Soon Tara and I had a date in Vancouver, and we both knew this was serious. On New Year's Eve, we hiked up Mount Hollyburn in North Vancouver with one of my students, his girlfriend, and another couple to stay in a cabin there. That night,

when we were in our sleeping bags, I asked Tara to marry me. And she did, on December 10, 1972, exactly a year to the day after we had met.

My children have been my pride and joy, but getting Tara to marry me was the greatest achievement of my life, and our marriage continues to be an adventure. Even now, when I come home from a long trip, my heart flutters at the thought of being with her.

When we were buying the house that is our home today, I had to ask Tara's parents for help to make a down payment on the mortgage and suggested that when Harry retired, I could add another storey to the house and they could come and live with us. We bought the house, added a storey to it later, and they moved into their separate apartment with us in 1980, an arrangement that worked wonderfully for all of us.

I was shocked one day when Tara told me that although she loved being with me and travelling to new places and meeting new people, she

LEFT: Parents-in-law Harry and Freddy Cullis on Freddy's eighty-sixth birthday.
RIGHT: Getting Tara to marry me was the greatest achievement of my life, and our marriage continues to be an adventure.

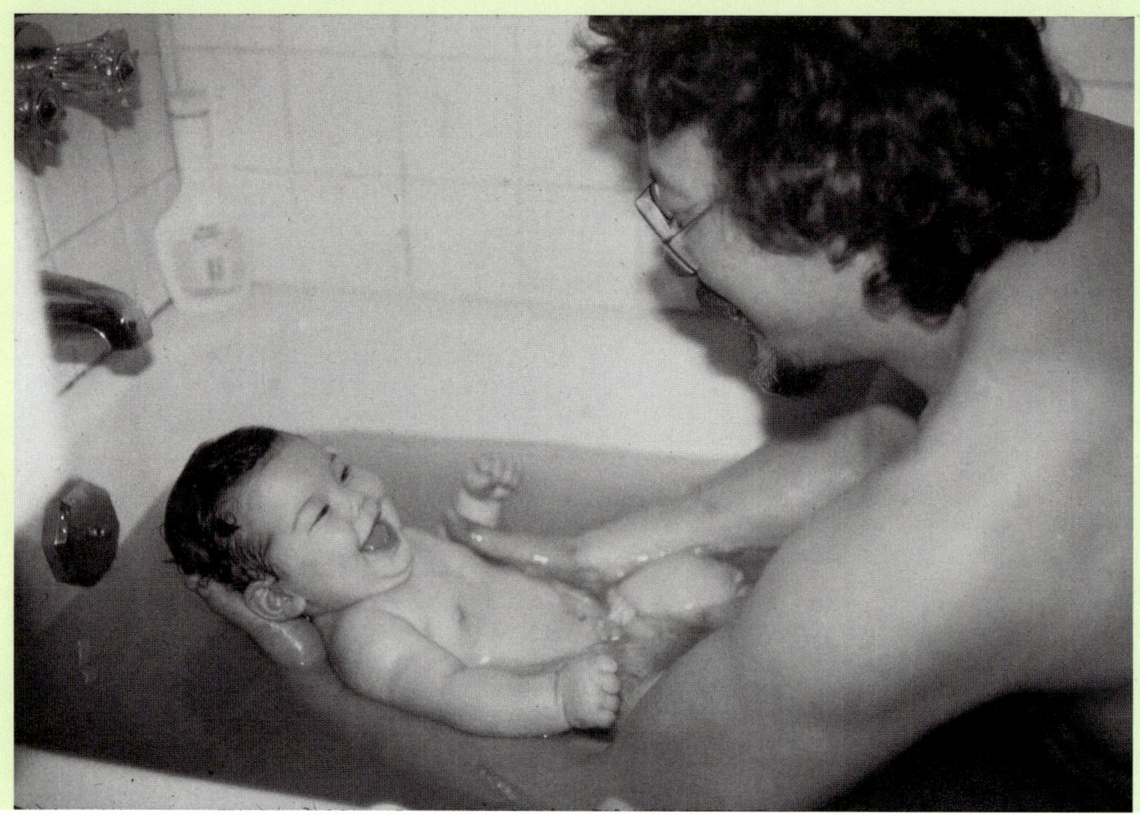

didn't just want to be my wife. She wanted to pursue studies beyond a master's degree. I was devastated but encouraged her not to apply to UBC but to schools with extensive programs in her field of interest, and she ended up being accepted at the University of Wisconsin in Madison. We had decided we would call each other every day, regardless of the cost. Those calls became our lifeline, something we continue to this day when we are apart.

My three children with Joane were a high priority to me, and Tara and I agreed we would like to have children together—just not right away. Instead of oral contraceptives, Tara chose to have an IUD. After I did an informative program on fertility in women, I told Tara she should have the IUD removed. But before we even sought a doctor, Tara found she

One of the great joys of parenthood, bathing Severn.

was pregnant. That bit of protoplasm climbed past the IUD and implanted in Tara's uterine wall. Severn arrived to the great joy of my parents, who were retired. During Tara's pregnancy, we had begun renovations on our house so that Tara's parents could move in with us, and Severn was their first grandchild, so they were thrilled. It had been sixteen years since the 1964 birth of my youngest child in my first family, Laura, so having Sev felt like starting anew. When Sarika arrived three and a half years later, Sev was running and talking and entertaining us with her cleverness.

In the meantime, Mom was beginning to show signs of forgetfulness. Dad and my sisters insisted she had Alzheimer's disease, but I denied it, because Mom exhibited no change in temperament. By the early '80s, though, it was clear she was losing her short-term memory.

On April 25, 1984, a month after they celebrated their fiftieth anniversary, Dad and Mom walked a few blocks to a local restaurant, had a meal together, and then went to a movie. As they were walking home, arm in arm, Mom had a massive heart attack and dropped to the sidewalk. She never recovered consciousness, though paramedics arrived quickly and got her heart going again.

I was in Toronto at the time and rushed home to be with her for the week before she finally died on May 2. As Dad said, "She had a good death." She didn't suffer, she was not incapacitated physically, and she had been with him right up to the heart attack. An autopsy revealed that she did indeed have the brain-tissue plaques characteristic of Alzheimer's.

Preparing for a crab feast with Sarika.

Tara Cullis

PRESIDENT AND CO-FOUNDER OF THE DAVID SUZUKI FOUNDATION

When I met David, he was thirty-five. That's fifty-five years ago. It was a leap of faith when we got engaged two weeks later. But that faith has never been betrayed.

David loved the life of a scientist. He loved his lab and his students. He was offered tenure and lab space at MIT, as well as at Berkeley and others. But he turned it all down. He loved Canada, despite what it had done to his family during the war. And he had a new goal. Although he kept teaching, he gradually left his beloved lab bench to do something more important.

David saw the future clearly and had to get the message out. He loved science and understood its power to change the world; he knew science had to be put into context, that it was too important to leave to scientists and politicians alone. He felt science and technology had to be understood by the public so that that the public could demand they be managed properly and could insist moral ethics be developed to use them wisely. Luckily, he turned out to be a natural at television and at radio, so he started *Suzuki on Science* on CBC TV, while he was still teaching at UBC, and *Quirks & Quarks* on CBC Radio. *Suzuki on Science* morphed into *Science Magazine* and eventually forty-four years of *The Nature of Things*—fifty-five years on TV.

I want to tell you that he has never let you down. He has never faltered in his efforts to help the public make good decisions. He has never stopped sounding the alarm and searching for solutions. His goal has always been for the public to be informed enough to have a say in science and technology and their uses. He feels science must be understood and handled carefully if climate change or habitat destruction or the decline of the oceans is not to overwhelm us.

He didn't do this work alone. His family helped him. The CBC helped him. *You* helped him. The public helped him. I've never seen anything like it.

All the thousands of volunteers over the years! The David Suzuki Foundation staff. (He often says "I'm going to the lab" when he heads off to work, and he loves the foundation staff like he loved his students.)

It's never ceased to amaze me how much people want David to reach his goals. I've come to realize it's because you and I share his goals. We see him as a means to achieve things we know are important. He symbolizes what a lot of Canadians want. The Japanese have a tradition they call National Living Treasures. I sometimes feel I'm married to a National Living Treasure.

Once I asked David how he has the courage to speak out. He said he's scared as hell, but he can't keep his mouth shut.

I wish more of us were like that.

David works harder than anyone. He's never late. He loves physical labour. He doesn't say "I told you so." His greatest wish would be to be wrong about the future we're headed into. There's still hope that nature will be "more forgiving than we deserve." David's kept the faith and so have you. Our goals and dreams are still intact, after all.

With Tara celebrating at a gala for my Legacy Lecture in 2009.

Ravi Jain and Miriam Fernandes

ARTISTIC DIRECTORS,
WHY NOT THEATRE

For almost twenty years, our company, Why Not Theatre, has been telling stories to bring people together in the hopes of making a difference in the world. We wanted to make a play about the planet. We thought if we wanted to tell a story that might inspire an audience to change our ways, who better than David Suzuki to do that with?

David said yes immediately, his curiosity about what we might do oozing out in each story he immersed us in. Stories of victories, depression, from more victories to "It's too late, we're not going to make it." The one character who was a constant in all his adventures was Tara Cullis, David's wife and collaborator of over fifty years. She was clearly the heart of the story and, we'd soon learn, the brains of the operation! So, Tara joined the project.

For three years, we sifted through their incredible adventures in places like the Amazon or Stein Valley and through endless albums of their photos documenting lifetimes! The story that emerged surprised us all. It was something deeper and more profound than "start recycling," "fly less," or even "do something small each day." The story was more personal, more vulnerable, and its ask was simple: Fall in love with the planet. It is a profound and fundamental shift in our understanding of what our relationship to the planet could be. When you love something, you will do anything you can to protect it. If we could love the planet the way we love those dearest to us, would we change? It was the question that guided our creative search with David and Tara, and spending time with the two of them, in the presence of their love, respect, and admiration for one another, the answer was clear.

David's life's work has been about getting the world to fall in love with nature. We saw it first-hand throughout our time with him; we saw how his love for the planet is as deeply felt as his love for his family, grandchildren, and, most of all, Tara. It is a privilege to be in the presence of a love like that, and we wanted to share it with the world through our play. David's love radiates still and is the force we need desperately to continue his life's work; to love this planet and to see ourselves in a reciprocal relationship with Earth, not just to take what we need but also to give back.

So, we made a play, *What You Won't Do for Love*. It continues to this day, and David's passion for it is, like him, unstoppable.

Perhaps it's the message of the play. Or maybe it's the connection that happens with the audience after the show, that thing only live theatre can do. Most probably, though, it's because Tara is the star. Her story shines bright, and their love for each other and the planet inspires audiences night after night in the most epic love story that encompasses us all.

Sturla Alvsvåg and Miriam Fernandes rehearsing the play *What You Won't Do for Love* with Tara and me.

It's A Matter
Of Survival

6 | HAIDA GWAII AND THE STEIN VALLEY

IN 1974, A GROUP OF CITIZENS on Haida Gwaii demanded protection of critical parts of the islands from clear-cut logging. In response, the British Columbia government set up the Environment and Land Use Committee. In 1979, one of the committee's recommendations was to not log in Windy Bay. That was not an acceptable option to the forestry company, which continued to press the B.C. government to allow logging.

I heard about this controversy in 1982, when I received a handwritten note from the New Democratic Party member of Parliament representing the Skeena riding, which includes Haida Gwaii. In his note, Jim Fulton wrote, "Soozook, you and *The Nature of Things* should do a program on Windy Bay." I could see it would be an important story, and I suggested to Jim Murray, executive producer of *The Nature of Things With David Suzuki*, that we do a program on the fight over its fate.

For the program, I interviewed Tom McMillan, then federal minister of the environment; environmentalist Thom Henley; Bill Dumont, with Western Forest Products; forester Keith Moore; researcher Nick Gessler; ecologist Bristol Foster; traditional Haida culture and language-holder Diane Brown; Miles Richardson, then president of the Haida Nation; and Guujaaw, a young Haida artist and carver. Ruggedly handsome, long hair

Severn and Sarika admiring a spawned-out chum salmon in the Stein Valley.

loosely braided, Guujaaw changed the way I viewed the world and set me on a radically different course of environmentalism.

I knew that unemployment in Skidegate and Masset, the two Haida communities, was high; that some of the loggers were Haida; and that the non-Haida forest workers often spent money in the two communities. One would think the Haida would welcome forest companies; yet Guujaaw had been a leader in opposing logging. When I asked him why, he answered, "Our people have determined that Windy Bay and other areas must be left in their natural condition so that we can keep our identity and pass it on to following generations. The forests, those oceans, are what keep us as Haida People today."

This was a statement of a fundamentally different relationship with the "environment" than most of us have, a sense that we are where we live, a relationship that is essential to future generations for whom Haida People feel a responsibility.

The Big Four on Haida Gwaii: me, Miles Richardson, Jim Fulton, and Alfie Collinson.

I continued my interview: "So if the trees are logged off—" Before I could finish my question, he responded, "If they're logged off, we'll probably end up the same as everyone else, I guess."

It was only days later, while I was watching the rushes, that I recognized the enormity of this insight. Since then, Guujaaw has confirmed that Haida People do not think they end at their skin or fingertips. Guujaaw opened for me a window into a radically different way of seeing the world.

The Nature of Things With David Suzuki program on Windy Bay was broadcast in 1982 to a large audience. After it ran, the South Moresby Resource Planning Team reached the same conclusion as the committee before it: Windy Bay, a jewel set in the misty isles, had to be protected from logging. Premier Bill Bennett did what politicians often do in such circumstances—he set up yet another group, the Wilderness Advisory Committee. After years of deliberation and no decisions about Haida Gwaii, Bennett set a ludicrously short time to decide on all these areas. Environmentalists immediately pointed out there wasn't enough time to perform the job responsibly, adding that the committee's membership was too heavily weighted toward the logging industry.

In the end, the committee came through, recommending that 363,000 acres, including Windy Bay, be set aside as parkland. Bennett was still under too much pressure from the forestry industry and loggers to accept that recommendation. At last, in 1987, new premier Bill Vander Zalm decided to include the disputed land in a park to be jointly administered by Parks Canada and the Haida People and known as Gwaii Haanas National Park Reserve and Haida Heritage Site.

I was invited to the provincial government buildings in Victoria for the July 1987 signing of the agreement between B.C. and Canada that would

Guujaaw has taught me a lot. From *Force of Nature: The David Suzuki Movie*

help to create Gwaii Haanas National Park Reserve. Later, we were flown across the water to Skidegate and were ushered into the village's great hall, where tables were set for a feast. A row of hereditary chiefs in full regalia presided over the long head table.

Allan Wilson, a Haida hereditary chief, leaped up at the Skidegate feast and publicly announced that he was giving me his dance apron, part of the formal regalia. Decorated with strips of copper, buttons, and figures of whales and birds, it was the first piece of regalia I ever received and is a much-treasured gift.

The Gwaii Haanas National Park Reserve and Haida Heritage Site agreement was signed in January 1993, after almost six years of negotiation between Canada and the Haida Nation. The official title recognizes that the Haida designated Gwaii Haanas a Haida Heritage Site in 1985.

Another battle that took place during this period was the fight to protect the Stein Valley; at 1,100 square kilometres, it was the last large unlogged watershed in southwestern B.C., relatively close to Vancouver.

In 1984, I received a request from organizer John McCandless on behalf of Chief Ruby Dunstan of the Lytton First Nation and Chief Leonard Andrew of the Lil'wat Nation to speak at the first of what was hoped would become an annual festival to celebrate and protect the Stein Valley. Unfortunately, I had a previous commitment during the inaugural festival in 1985 and couldn't make it, but I attended the next year.

John had conceived the idea of raising the Stein Valley's profile by holding a First Nations–run festival that would feature speakers and musicians. That first gathering attracted up to five hundred people, who hiked high into the alpine at the valley headwaters in a terrific kickoff to what would become an ongoing success.

For the second festival, Tara and I were delighted to have the chance to camp in a part of the province we hadn't seen before. The festival site was alongside the lower Stein River in a meadow, and there might have been a couple of hundred people there. In preparing for my talk, I had to integrate my ideas about the environment with what little I knew about the traditional values of First Nations.

In the next tepee were Miles Richardson, the charismatic young president of the Haida Nation, involved in his own battle over the land; Patricia Kelly, his Coast Salish girlfriend; and Guujaaw, the Haida artist who played such an important role in my education and who would himself become Haida Nation president after leading the fight against logging in Haida Gwaii. They would become our dearest friends and companions over the years.

Thanks to increasing attendance at the festival, interest in the Stein Valley grew. The environmental community rallied to the cause. I called and recruited the Canadian singer Gordon Lightfoot, who flew his entire band to the Stein to perform for free. Later, Gordon became a very good

LEFT: **Sev, Guujaaw, and Tara at a Stein Valley Festival in the alpine meadows.**
RIGHT: **John Denver and his wife, Cassandra Delaney, at the 1987 Stein Festival.**

friend and donated a large sum of money that pulled the festival out of debt. Gordon died in 2023.

In 1987, I was able to find a phone number and call the American singer John Denver, who surprisingly knew of me and accepted my invitation to perform at the Stein and, like Lightfoot, travelled in at his own expense. John became a friend. I'm glad he knew before he died in 1997 that the Stein Valley had been set aside as a provincial park.

In October 1987, B.C. Minister of Forests Dave Parker gave the go-ahead to log the Stein. However, the buildup of support for protecting the valley paid off: Because of the festivals, the Stein had become too well known and support for its protection too great to send the loggers in.

By 1988, 3,500 people were attending the Stein Valley Festival. The following year, 16,000 people drove to the event, held at the rodeo grounds near Mount Currie. They were entertained by Canadian stars Bruce Cockburn, Gordon Lightfoot, Colin James, Valdy, Blue Rodeo, and Spirit of the West, among others. The festival had become so huge that it now made money, and I was sure the size of the crowd ensured the valley would never be logged.

In 1995, B.C. Premier Mike Harcourt held a ceremony with Chief Ruby Dunstan and Chief Leonard Andrew to set aside the entire watershed as Stein Valley Nlaka'pamux Heritage Park, administered by the Lytton First Nation and B.C. Parks.

In Haida Gwaii and the Stein Valley, the battle was led by First Nations. Theirs was a struggle over land, which they knew they had a responsibility to care for, not for the superficial needs of money, jobs, or control, but for the most powerful need of all—to remain who they are.

Guujaaw

ARTIST, SINGER, HEREDITARY LEADER AND FORMER PRESIDENT, HAIDA NATION

It's hard to think of David as normal, having a household name,
the familiar voice from many a generation's childhood
and the fearless face we all know so well
... none of that would be associated with "normal."
We know him as a great advocate for the planet,
deserving the title of a national and even planetary hero
... or to many the harbinger of bad news.
His cynical yet so logical rants are eaten up by his fans and detested by the bad guys.
He gets fired up when his detractors go after him
... he can be a wily son of a gun.
Our guy has been so disappointed by humans and mostly the governments,
... though he fights on.
and yet he thinks he failed
because in spite of his efforts,
they keep spewing out the poisons and wrecking the planet,
... not you, David, humanity has failed.
Our David is foremost a husband, dad, and grampa.
... he loves his family, good food, and he swears.
That's who he is, not just the clever and golden-voiced orator.
We are lucky to have David in the world with us.

Miles Richardson

OFFICER OF THE ORDER OF CANADA, DAVID SUZUKI FOUNDATION BOARD MEMBER, PARTNER AT RICHARDSON STRATEGY GROUP

We are celebrating a life well lived, a person who made a positive difference in a lot of lives. I am one of those people.

I met David in the early 1980s when he came to Haida Gwaii as host of CBC's *The Nature of Things* for a story about protests against logging on South Moresby (Gwaii Haanas)—Canada's largest environmental campaign. He interviewed me as the young president of the Haida Nation. My primary task was to turn this huge campaign into a cutting-edge Haida title assertion. David generously gave our voice a platform and helped us tell the story of our ancient Nation to the world. He listened and strived to understand a fundamental tenet of Indigenous cultures around the world: All of creation is connected; we are all one. He has the courage to live his life respecting this truth.

We deepened our friendship by joining in support of various Indigenous Peoples who were protecting their homelands and upholding their stewardship responsibilities to their places: in B.C., Canada, Turtle Island, and around the world. David and Tara founded the David Suzuki Foundation with like-minded friends and supporters to respond to people's persistent question: "We believe humankind is facing an ecological crisis; now what do we do about it?" Solutions always began with respecting Indigenous nationhood and recognizing Indigenous title. David's voice remains one of the most trusted in Canada on environmental issues. We led a lot of progressive change.

These pursuits took us all over the world. While we worked hard and passionately, we had a lot of fun too. We met a lot of people and

experienced different cultures, always exploring and savouring our differences while shining a light on what we had in common.

David spends a lot of time in heavy analysis and reflection, but he also sees and very much appreciates the lighter side of the life we have—whether hanging out and sharing a good meal with family and friends; paddling down the Xingu River in the Amazon rainforest; fishing for salmon in the waters of Haida Gwaii, for tucunaré with Paulinho Paiakan and the Kaiapo, or for lake trout in the Sahtu; or just sitting around the fire with friends. David and his family really appreciate the simple, real things in life. David and Tara and their family have become much more than dear friends; they are my family too.

Miles Richardson, Tara, and me at the 2017 David Suzuki Foundation fellowship reception. Shannon Ruth Dionne Miller

7 | ADVENTURES IN THE AMAZON

WHEN I WAS A BOY, I would sneak a peek at Dad's adventure magazines, which carried tales of true-life journeys to exotic places. The ones that would make my heart beat wildly described the Amazon, a place I yearned to visit.

In 1988, at the age of fifty-two, I had my chance to realize my boyhood dreams. In August, *The Nature of Things With David Suzuki* crew travelled to Brazil to begin filming for a special program on the Amazon rainforest's ecosystem. A month later, I flew to the outpost of Porto Velho, capital of the state of Rondônia, to hook up with the crew. But my first glimpse of the legendary forests was bittersweet: We were there to bear witness to its destruction.

When I caught up with the crew in Rondônia, they had not been able to take aerial shots because smoke from the burning forest was so thick it was too hazardous for airplanes to take off. I was excited to be there, depressing as the scene was, but the team was demoralized by what they had filmed: poverty, malnutrition, malaria, and children so painfully thin the crew ended up giving them money for medicine and food. We filmed endless scenes of burning—trees, fields, whole forests going up in smoke.

Over my decades in television, I've learned that filming in another country can be a huge hassle. When the crew comprises host, producer, researcher/writer, cameraperson, camera assistant, soundperson, and

With Paiakan in A'Ukre.

lighting person, along with forty heavy bags of luggage (some metal trunks), a tough and savvy local agent is required to organize it all.

In Brazil, that was Juneia Mallus. When we said we needed to film an Indigenous person who could articulate the importance of the forest and show us through their community, Juneia knew who it should be: an extraordinary man she had worked with before—Paiakan.

We were to meet Paiakan in the Kaiapo village of Gorotire, which was once reachable only by trails but now had a road from the outside. As difficult as it was to manoeuvre on that road, it was nevertheless the opening for the influx of "civilized" products—white bread, candy, beer, liquor, tobacco—that pollute the community we were approaching. Disgusted with what that road had done to this village, Paiakan pulled out and moved far into the forest to establish a new village where his people could continue to live traditionally. He found the perfect place on a low bluff overlooking a river filled with fish. He called the community A'Ukre (Ah-oo-cray), apparently named after the sound a certain fish makes when caught. About two hundred people had decided to follow Paiakan and live in A'Ukre. But he was to meet us in Gorotire.

It was early evening, and we were relaxing in a hut in the village when Paiakan came by. He was controlled when he met us—not suspicious but curious. In Gorotire, our conversation had to be funnelled through Juneia in Portuguese, the official language of Brazil, which Paiakan had acquired as a teenager and now spoke fluently. Paiakan's father, Chikiri, was a chief. For his first fourteen years, Paiakan had lived a totally traditional life, as his ancestors had done for thousands of years, hunting and gathering according to knowledge acquired and passed on for many generations.

But even the immensity of the Amazon rainforest was not enough to protect the Kaiapo from the encroachments of the *brancos* (Europeans).

Paiakan.

The Kaiapo could smell the fires and were beginning to see extensive gold-mining pollution in some of the big rivers. Paiakan realized he had to learn more about the encroachers. At seventeen, he went to the Catholic mission, where he learned Portuguese and some Brazilian culture. After he learned to write, he promptly wrote a book about the forest as his home. Paiakan could have moved to a city and become an "urban Indian," but he wanted to learn enough to protect the traditional ways, and he moved back to his village. Paiakan became the acknowledged leader of the community.

The camera crew were to fly into A'Ukre in a couple of days, but Paiakan wanted to fly back to A'Ukre with his wife, Irekran, before the rest of the team. The extra flight would cost several hundred dollars. I offered to pay the money and asked if I could go in with them.

The next day, Paiakan, Irekran, and I took off for A'Ukre. As the plane levelled off, I had a moment of panic as I realized I would be spending the next days in a village where only Paiakan spoke any language other than Kaiapo and that was Portuguese, which I didn't understand. He had learned one English phrase: "Let's go, Dave."

My panic quickly passed, however, as I was swept up in the wonder of the adventure, my childhood dream come true. After an hour's flight over endless, pristine forest, a clearing came into view. I saw an oval ring of huts by a stream, the Riozinho, "little river." A thin cleared track was our runway. We bounced along the stubble to a halt, and the plane was mobbed by what seemed like the entire village.

That night I lay in my hammock, listening to the steady thrum of insects and the chirps of frogs from the surrounding forest, the gentle snores and breathing of Paiakan's family all around me. I felt so far away from anything I knew. This was the realization of dreams I had held for forty years.

Kaiapo girls in A'Ukre before a *festa*.

We interviewed Paiakan on camera, asking him why he had moved his people here and what the forest meant to him, as Juneia translated his Portuguese for us. He was eloquent, and it was a productive shoot. Then Paiakan sat down with Juneia to talk to me about his plans to fight the dams. He asked me for help in raising money to take tribe members to Altamira and to build the traditional village on the dam site. I had no choice but to promise I'd do the best I could. But if I were to raise funds, I realized a key question was: Would he be willing to come to Canada himself?

Soon we were on our way out of the village, crossing a sea of green that extended as far as we could see on both sides of the plane. As soon as I could, I phoned Tara. I related the threats to the forest. I told her about the Kaiapo and their charismatic leader, describing Paiakan's plan, his need for funds, and his promise to come to Canada to help raise money and the profile of the issues.

As I continued for the remaining five weeks of the shoot, Tara sprang into action in Canada, organizing events in Toronto and Ottawa. In 1988, the Amazon was a hot topic. The scale of its destruction was on everyone's lips.

People were quick to lend a hand. In Toronto, Monte Hummel and the World Wildlife Fund offered support for a fundraising event, and in Ottawa, Elizabeth May, who was with the Sierra Club and had first rocketed into prominence fighting pesticide spraying in Cape Breton forests, promised the same. Soon great plans were afoot.

In Brazil in 2017 for the twenty-fifth anniversary of the 1992 Earth Summit. L-R: Severn Cullis-Suzuki; Tara Cullis; Paiakan's wife, Irekran (holding baby Tiisaan Brown); me; Paiakan's daughter Maial; Barbara Pyle; Paiakan.

Katsi'tsakwas Ellen Gabriel

KANEHSATÀ:KE NATION,
KANIEN'KEHÁ:KA (TURTLE CLAN),
ARTIST AND ACTIVIST

I met David in Tokyo, Japan, in 1992 for a conference at Meiji Gakuin University. Keibo Oiwa, a mutual friend, organized a conference commemorating 1492: five hundred years of "contact" in the Americas. Keibo Oiwa is an anthropologist who worked at Concordia University and was back to his home in Japan after 1990.

The three-day conference was an amazing opportunity to meet other Indigenous people and visit a land that has so much history and culture. I met David at breakfast one morning. He came to greet us all and sat to speak with us.

After the conference, we all travelled to Hokkaido, following David like we were his entourage. The national television station was doing a documentary about David, so we were able to witness the filming. We listened in awe as he spoke about the land, its biodiversity, and the consumerism of society. We shared stories and got to know each another as we gazed in amazement of the beautiful landscape of rolling mountains, as if from a Japanese watercolour. We all got to know David during this time and were amazed by his knowledge of the science of the land. His down-to-earth manner made him approachable, and we got to know some of his sense of humour. We also witnessed the depth of his love for the natural world, which he has advocated for over many decades.

A year later, David visited my community of Kanehsatà:ke, and the whole community was excited.

Niawenkó:wa—a big thank-you for advocating for Mother Earth and all our relations.

Keibo Oiwa

CULTURAL ANTHROPOLOGIST, ACTIVIST, AUTHOR, PROFESSOR EMERITUS AT MEIJI GAKUIN UNIVERSITY

In 1977, I moved from Japan to North America. While studying at McGill University in Montreal, I heard about David Suzuki. I became involved in research on Japanese Canadian history and participated in the movement to seek apology and reparation for the Japanese Canadians who had been persecuted during the Second World War. I met David when he came to Montreal to speak in support of the redress movement.

In that speech, he said something that moved me: "The reason Japanese Canadian redress is meaningful to me is because I believe it is going to be a step toward the more important and fundamental issue for this society: redress for First Nations people and establishment of Indigenous rights."

In 1991, I returned to Japan with my family to teach at a university in Yokohama. There, I hosted an international forum, "Native Wisdom for the 21st Century: 500 Years Since Columbus—Another Vision." Indigenous leaders from around the world presented an alternative perspective for the 500th anniversary of Columbus's arrival in the Americas in 1492. The textbook that guided me was David's new book, *Wisdom of the Elders*.

David accepted my request to participate as a keynote speaker. After the conference, I led a tour to Ainu communities in Hokkaido with David and others. With NHK Sapporo, we had the trip filmed for a TV program. I titled this thirty-minute mini-documentary *In Search of the Lost Forest*.

Near the end of David's visit, I suggested he compile a book titled *The Japan I Never Knew*. David responded, "You kidding? I might look like one, but I am not Japanese. I don't speak the language and

am totally ignorant about Japan. How could I write a book?" The next day, David said to me, "About your idea of the book, I think it would be great if you would co-author it with me." My reaction was akin to David's reaction the night before: "Are you kidding me? I fled to North America and just returned from living abroad for fourteen years. How could I write a book about Japan?" David and I eventually co-authored *The Japan We Never Knew*.

Since my Japanese translation of David's *You Are the Earth* was published in 2007, I have used it as a textbook for first-year students. My translation of David's 2009 Legacy Lecture was published the following year. During that time, David visited Japan several times and gave lectures. David and Tara's daughter Severn also came to Japan several times at the invitation of the Sloth Club, which I founded twenty-five years ago with friends and students.

In summer 2024, I visited Canada for the first time in years. David and Tara invited me, my daughter, and my five-year-old grandson to their cottage. David took my grandson, who grew up in the city, on "adventures" with his grandchildren at the beach, in the bush, and in the swamp. This was a turning point in his life. David's magic is alive and well.

With Keibo Oiwa.

8 | PROTECTING PAIAKAN'S FOREST HOME

IN FEBRUARY 1989, we had arranged for Paiakan to fly to Toronto for our concert to raise funds for the protest to be staged at Altamira. Tara had an audacious idea: Why not invite the major multinational companies that did business in the Amazon to attend a reception before the concert to meet Paiakan in person and, in return, to donate $1,000? We drew up a list of eighteen companies, from American Express to the Bank of Japan. The CBC filmed the arrival of the hosts—me, Paiakan, the Canadian writer Margaret Atwood, and Gordon Lightfoot—and the well-heeled guests. Of the eighteen companies, all but one—the Bank of Japan—sent a representative with a cheque. In one hour, we raised $17,000.

The main event was a concert at St. Paul's Church on Bloor Street. More than three thousand people jammed that church. Margaret Atwood read a poem and Gordon Lightfoot and a hot a cappella group, the Nylons, sang.

Paiakan appeared onstage in a shirt and pants, but his face was painted and he wore a brilliantly coloured feather headdress. The grand room went silent as he spoke about his forest home, which had supported his people for so long, the threat the dam posed, and his need for our help. When it was over, we had raised more than $50,000. We went on the

Paiakan and I display his wife Irekran's paint job.

next day to Ottawa and another gala event. Elizabeth May gave a brilliant speech, and once more Gordon Lightfoot performed. By the time Paiakan left, after only a couple of days in Canada, we had raised $70,000.

The Nature of Things program was broadcast as a two-hour special entitled "Amazonia—The Road to the End of the Forest," and it garnered a huge audience. Tara began the difficult task of arranging our trip to Altamira, a frontier town deep in the Xingu valley of the Amazon. Tara learned some Portuguese as more and more enviros called to ask to go along with us. She managed the heroic job of booking planes and hotels for forty people!

We had a virtual who's who of the Canadian environmental movement travelling with us, including Elizabeth May of the Sierra Club, Peggy Dover of the World Wildlife Fund, Paul Watson of the Sea Shepherd Conservation Society, Jeff Gibbs of the Environmental Youth Alliance, Peggy Hallward of Energy Probe, Gordon Lightfoot making good on his promise, Guujaaw of Haida Gwaii, and Simon Dick, a Kwicksutaineuk from Kingcome Inlet.

Paiakan says farewell to Tara and me at the Altamira airport.

TOP: Paiakan (far right) lifts an electric eel while Kaiapo youth and helper Mokuka dispatches another one. That's me on the top left, with another Kaiapo youth, Caro. RIGHT: News conference at Altamira. That's Sting standing, Paiakan seated looking up, and Simon Dick in full regalia behind him.

From Manaus, we flew to Belém. En route we spied a newspaper with a picture of Paiakan in a hospital bed! Tara grabbed the paper and learned that Paiakan had appendicitis; he still intended to appear.

To our relief, he was there at the opening—grey and weak but still clearly the leader. It was a huge event, with some six hundred Kaiapo representatives and forty other tribes all seated around the dais wearing little more than shorts, feather headdresses, and body paints; there were hundreds of Brazilians from officials at Eletronorte, the power company proposing to build the dam, to FUNAI, the Brazilian government organization for Indigenous Peoples; environmentalists and reporters from around the world to hundreds of armed soldiers ringing the walls inside the building.

Under pressure from many countries, including Canada, the World Bank pulled its support from Plano 2010, bringing the hydro power project to an end. After the meeting, Paiakan took off his headdress, which his mother had made for him, and gave it to me. It's another of my most prized possessions.

While in Altamira, Tara and I met late at night with a handful of trusted Brazilians concerned that Paiakan's life was in jeopardy. Indigenous people are murdered, and the killers often receive no consequences. Paiakan had received several death threats. He needed to get out of the country. We asked him where he would like to go. "To Canada, to stay with you and Tara," he replied. Within days of our return to Vancouver in March 1989, Paiakan arrived with Irekran and their three daughters,

Demonstrating the headdress given to me by Paiakan at Altamira.

Oe, Tania, and baby Maial. Irekran and the girls spoke only Kaiapo, so our communication was through Tara and Paiakan speaking Portuguese.

We took the family to visit as many different First Nations as we could. We visited Tofino, on Vancouver Island, where the Nuu-chah-nulth People were holding a meeting. Paiakan was feted like a relative. We took him to Alert Bay, home of the Kwakwaka'wakw People. When we arrived at the ferry terminal in Alert Bay, we were met by Kwakwaka'wakw dancers in full regalia. We travelled to Haida Gwaii, where Paiakan was taken out on *Lootaas*, the canoe that was paddled from Vancouver to Haida Gwaii during the battle to save South Moresby.

I decided to return to A'Ukre years later to see Paiakan while we were in Brazil filming for *The Sacred Balance*. Paiakan was heavier, and the village, too, had changed since my last visit. For some unfathomable

LEFT: Sarika shows where the parasitic worm was in her foot. RIGHT: Paiakan's daughters Oe (left) and Tania (right) with Sarika playing dress-up in the shower.

reason, the thatched roofs had been replaced with metal. A dispensary with a concrete floor had appeared, staffed by a Brazilian who gave out medical drugs; a solar-charged television set was turned on for a few hours a night to show soccer while I was there, and a hut had been built for people who were coming and going to the research station upriver. In A'Ukre, I woke to the tap, tap, tap of metal devices being used to shell Brazil nuts for the Body Shop chain, which uses the extracted oil in its cosmetics. The plane we had delivered in 1989 still linked the Kaiapo villages.

The day I left A'Ukre, I was wakened by a horrendous racket, which I learned was the pharmacist spraying insecticide around the village—malaria had come to this part of the forest. It seems there is no way to escape the forces of change, even in the deepest part of the Amazon.

Sadly, Paiakan died of COVID-19 early in the pandemic, in June 2020. His daughter Maial was able to realize their dream of her attending university. She has since become a committed communicator and activist for Indigenous rights. I met her again at the 2019 COP25 climate summit in Spain. She had come to Vancouver as a baby and now was fulfilling her father's hope that she would carry on the battle to protect their land—though the dam we stopped in 1989 has since been built. She was awarded a Suzuki fellowship from the David Suzuki Foundation the following year and invited me and the family to visit her village again. Tara and Severn hope to go now that Maial has had a baby.

Severn with Iremaõ, Paiakan's son, at the Pinkaiti research station. Jeff Topham

Margaret Atwood

AUTHOR, POET

It's hard to believe that David Suzuki is ninety. Surely, he is timeless. It seems he has always been with us, exploring, explaining, wondering, revealing, and warning. He has helped us to know and love our planet and our biosphere, to marvel at its infinite variety, its strangeness, and its familiarity; to appreciate our place in it; and to realize that if our home is destroyed, we ourselves will vanish. His influence on many generations has been profound; his wisdom is enduring. It's an honour to know him.

Margaret Atwood reads my palm on Granville Island in 2023. Tara Cullis

Elizabeth May

LEADER OF THE
GREEN PARTY OF CANADA,
MEMBER OF PARLIAMENT

I met David when I was a campaigning anti-pesticide waitress in Cape Breton in the 1970s. He was in Halifax, and I phoned every hotel trying to reach him. After half a dozen calls, I finally succeeded, but his response was "Who is this? You just got me out of the shower!" I stammered through asking my hero if he could please forgive me. I gave the pitch on how we needed his help to stop aerial spraying of organophosphate insecticides over our forests. He said, "Yes."

We have since worked together on so many issues—from helping end logging on Haida Gwaii, to working to stop pipelines, to protests and rallies too many to count where just hearing David Suzuki speak in person brought people to tears.

My favourite story is so unlikely—we were in the Amazon where we had travelled to support Kaiapo leader Paiakan, whom David had met while filming *The Nature of Things*. Learning of Paiakan's fight against a hydro dam on the Xingu River, David had come home and recruited Tara and me, and we recruited others to raise money to fund Paiakan's campaign plan for a pan-Amazonian gathering in Altamira of Indigenous people to protest the dam. It would be the first time people from so many tribes from all over the Amazon had gathered in one place.

In February 1989, the whole mad scheme had come together! Thanks to Gordon Lightfoot doing concerts in Ottawa, Toronto, and Vancouver, we raised the money. Then, impulsively, Tara said she and I should attend. The whole thing snowballed to about forty Canadians, including David and Gordon, Tara and me, and a rogue's gallery of eco-activists, from Sea Shepherd captain Paul Watson to youth leader Jeff Gibbs. David introduced me to another friend of his, Yanomami leader Davi Yanomami. The conversation was not fluid, going from English to Portuguese to Yanomami and back to English.

Davi Yanomami thought I was David's wife. When that got translated, David was horrified. "No, not my wife!" But then in a sweet effort to make me feel less rejected, he said, "But if I could have two wives, I would have her too." Davi Yanomami expressed concern: "Oh no, two wives under same roof is a very bad idea."

Tara and I decided it could be good. I never mind washing dishes but hate ironing, and she has the reverse preferences. And both of us love David—but quite differently, I hasten to add! David Suzuki is (as we all know) a genius, and the smartest thing he ever did was marry Tara!

What few among his vast following of admirers may know is how interested he is in everybody else's romances and love lives. He was a consistently reliable shoulder to cry on whenever a relationship came to a crushing end and a great cheering section when I married John Kidder, whom he and Tara had also known for ages!

Our lives are well mixed—daughters and in-laws, grandkids and partners, all woven together in a consistent commitment to fight against all odds to protect life and love and truth.

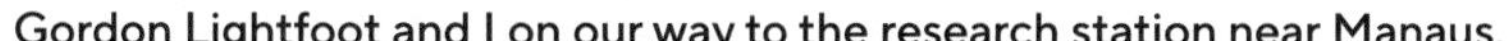

Gordon Lightfoot and I on our way to the research station near Manaus.

9 | DOWN UNDER

AS A BOY, I dreamed of visiting the Serengeti in Africa, the Amazon in South America, and Australia where I longed to see a duck-billed platypus. I had received a job offer from the University of Sydney in the early 1970s but was reluctant because of the country's "whites only" immigration policy. In 1988, I did go to Australia and immediately fell in love with it. Contrary to my expectations, Melbourne was a vibrant metropolis with a sizable Asian population. The environmental movement and public support around the world were at a peak, and Australia had recently established the Commission for the Future, a government-funded organization that looked at the role of science in Australian society and its place in the country's future.

Phil Noyce, a science teacher who was recruited to the Commission for the Future because of his interest in communicating science to the public, had encouraged the organization to invite me to give a series of talks in 1988. He would later become a close friend who convinced me of the importance of acting immediately to fight climate change. (Tragically, he had an undetected congenital heart defect and died in the prime of his life, while playing tennis.)

Quokkas, a type of marsupial, on Rottnest Island, near Perth, Western Australia.

As a result of my first visit, several groups invited me to return to give talks, and I was determined to go back with Tara. I was also approached by Patrick Gallagher, head of Allen & Unwin Book Publishers. When I told him I wanted to set aside some of my Australian book royalties to support Aboriginal and environmental groups in Australia, Patrick promptly offered to contribute 5 percent of the profits the company makes on my books to the fund.

Soon I was planning a return visit to Australia, this time to tour several cities to talk about my books *Metamorphosis* (my first autobiography) and *Inventing the Future* (a collection of columns I had written for newspapers). In 1989, Tara accompanied me. Sales of the books took off, and they became bestsellers.

On that visit in 1989 with Tara, Phil Noyce and his wife, Georgina Tsolidas, accompanied us on a jaunt to the Great Barrier Reef. When we returned from our enchanting trip, we noticed an excavation site a block away from Port Douglas's Four Mile Beach with a sign saying an apartment building was soon going up and one suite was still for sale. Phil, Georgie, Tara, and I bought it. Tara and I were able to share the suite with friends. But I am a carbon sinner, having flown extensively for *The Nature of Things*. Since I cut back on flying, trips to Australia were no longer feasible so we reluctantly sold our part of the suite to Georgina.

Before my third visit to Australia, I received a message that Peter Garrett, who sang with Australian group Midnight Oil, had offered to do an event with me in Sydney on my next visit. I wrote to ask Patrick Gallagher who this guy was and whether he was legit. Patrick replied that he was a big name and Midnight Oil was a very popular group, but the publisher seemed hesitant about the idea of my doing an event with a rock band. I told Patrick to thank Peter but turn him down. I soon learned that Peter and the Oils were not just big in Australia, they were huge!

Peter was president of the Australian Conservation Foundation, one of the largest environmental groups in the country, and he had run

unsuccessfully for the Australian Senate in 1984 for the Nuclear Disarmament Party, a predecessor of the Greens. He was more successful with the Australian Labor Party, serving from 2004 to 2013, including as environment minister and education minister. When I heard their signature song, "Beds Are Burning," demanding that Australians confront the fact that the land belongs to the Aboriginal Peoples, I was blown away.

I saw him onstage for the first time in Anaheim, California, and his rapturous reception by the audience showed me how badly Patrick and I had erred, but by then Peter and I had become good friends. When he came to Canada in 1993 as the long battle over logging in the Clayoquot Sound rainforest on the west coast of Vancouver Island was heating up again, Midnight Oil volunteered to perform in the protest area, and I shared the stage with them.

On another early visit to Australia, I received a request that I meet Green Party politician Bob Brown of the state of Tasmania. He was a

I took this photo of Aboriginal people in Arnhem Land performing for *The Sacred Balance*.

doctor who loved the forests of Tasmania and became an environmental activist against the logging industry. Eventually, he went into politics. As an elected senator in the federal government, he played an indispensable role in raising the profile of environmental and human rights issues. Bob wanted to know whether Lake Pedder, a pristine glacial lake in Tasmania's southwest wilderness that had been flooded in the 1970s to generate hydroelectric power, could be restored. To study such matters, he needed money. It was the early days of the David Suzuki Foundation, and I was pleased to be able to support an international project under our name.

One of our most memorable visits to Australia occurred in 1991, when my father had recovered from my mother's death in 1984 and had regained some of his great zest for life. Dad adored our children, and we invited him to join us on a trip to Australia.

With Georgina and Phil in tow, we made our way to Port Douglas. There Tara bought an inflatable vest for Dad, and we took the family out to the Great Barrier Reef. We fitted him with mask, fins, and snorkel to go along with his protective vest, and in he jumped, hand in hand with Sarika. It is one of my happiest memories.

When Mike Rann was minister of youth affairs and of Aboriginal affairs in the state of South Australia's Labor government in the 1990s, he asked me to be the honorary patron of the Youth Conservation Corps (YCC), a program that gave young people a stipend to spend six months a year learning how to rehabilitate the land. But when Labor was thrown out of office in 1993, the program was cancelled. Mike led the party back to power in 2001 and as state premier then resurrected the Youth Conservation Corps and asked me to return as honorary patron. I was delighted, and Tara and I attended a YCC event in Adelaide.

We were taken to a large area of degraded land where the trees had long been cut down, the land overgrazed by sheep, and the soil overrun by grass and brush. "This will be Suzuki Forest, named in honor of our patron," announced John Hill, minister for environment and conservation.

In Adelaide after one of my readings, Lewis O'Brien, a respected elder of the Kaurna People, approached me and introduced himself. He was pleased because I had talked about the book I co-authored in 1992 with Peter Knudtson, *Wisdom of the Elders*, which examined the congruence between Indigenous knowledge and scientific insights.

In a simple gesture, Lewis said, "I want to give you a name—Karnemeyu. It means 'holy mountain.'" Receiving a name from Indigenous Peoples is the highest honour I can imagine. What surprises me is that of the names I have received from Indigenous Peoples, three have meant "mountain." Simon Lucas, Nuu-chah-nulth from Ahousat on Vancouver Island, gave me my first such name, Nuchi, meaning "mountain," and the Kainai Nation (previously known as the Blood Tribe) near Lethbridge, Alberta, gave me the name Natooeestuk, meaning "sacred mountain."

The history of Australia over the past two centuries has been one of decimation of the Aboriginal Peoples, a deliberate attempt to eliminate them by killing or through assimilation. A climate of racism has led to

LEFT: A party greeting me at the airport in Port Moresby, Papua New Guinea. RIGHT: Dressed up to entertain.

enormous problems for the survivors. But as the twentieth century ended, Australians overwhelmingly wanted justice and reconciliation for Aboriginal Peoples.

On several visits to Australia, I was invited to Papua New Guinea, where isolation imposed by the rugged terrain has resulted in a profusion of cultures and over seven hundred languages, about 45 percent of the world's total. I first visited in 1992, at the invitation of Nick Fogg, who worked for CUSO (formerly Canadian University Services Overseas and still known by the old acronym), an international developmental NGO.

Nick had arranged for a visit to a series of remote villages, which we reached by boat, plane, or jeep. Each village I visited entertained me with performances and plays the people had made. I was treated generously, feted, and fed traditional foods.

Trying out the drum given to me by people in the Owen Stanley Range in Papua New Guinea.

Like the Kaiapo in the Amazon, the people in these remote villages were almost completely self-sufficient. Nevertheless, the products of the industrial economy were visible in every Papua New Guinean village I saw. Having become an independent nation, Papua New Guinea must find a source of revenue for the government and bring the benefits of education and medical care to very remote communities. The challenge is to decide whether traditional customs and practices have value in a global economy and, if they do, how to protect them while generating income.

I returned in 1994, invited by women in a town called Wewak in East Sepik province. They were concerned about the loss of their forests and rivers.

I was intrigued by Papua New Guinea, because it has one of the largest intact tropical rainforests left on the planet and is still occupied by the Indigenous people who own the forests by law. I promised to help by sending money from a fund I had set up in Australia from the profits of my books there.

With William Takaku, environmental activist, artist, and actor, who starred as Man Friday in the movie *Robinson Crusoe* with Pierce Brosnan.

Bob Brown

AUSTRALIAN POLITICIAN, MEDICAL DOCTOR, ENVIRONMENTAL AND HUMAN RIGHTS ACTIVIST

David Suzuki's first speaking tour of Australia, in 1988, amazed us all. The University of Tasmania's largest theatre was packed. We were a society mesmerized by the big lie of endless economic growth, but David declared that we had to ditch growth economics by putting Earth's ecosystem first. How could anything so outrageously contrary to our materialist culture be uttered in this island's citadel of learning? It left many listeners reeling at first, but here was a thoughtful audience and, as if a moment of monumental truth had dawned, the crowd began clapping and then foot-stamping in a burst of excited and unrestrained applause.

This visitor from Canada had freed us environmentalists Down Under from an intellectual underground. Thanks to David, the logic of prioritizing life on Earth had become legitimate as never before.

In 1992, the David Suzuki Foundation gave initial support to our campaign to recover Tasmania's beautiful Lake Pedder and its highland beach and sand dunes, which had been inundated by a hydroelectric scheme in 1972 to produce a pitiful sixty megawatts of electricity. The foundation funded a study that demonstrated the economic, geological, and biological feasibility of restoring the lake. The Restore Lake Pedder campaign continues to grow, is now on the national agenda, and will yet succeed.

Always welcomed by environmental and Indigenous communities, David also ran into the cynics of science and corporate hucksters who are in abundance in Australia.

In 2006 at the National Press Club in Canberra, David warned us, in graphic fashion, about the environmental disaster of the nascent

Atlantic salmon farming industry. Before his speech, the luncheon was served: grilled Atlantic salmon. The guest of honour had it sent back to the kitchen and gave his speech on an empty stomach!

Now, there's a national uproar over the Atlantic salmon industry, owned by multinational corporations and polluting Tasmania's inshore waters. In 2025, a bacterium got into the overstocked cages, and despite the tonnes of antibiotics added to the marine environment, millions of the caged fish died and globules of their fat washed up along miles of the island's scenic beaches. Thousands of people protested on the beaches. We can't say we weren't warned: David had made an enduring statement against an industry that, two decades later, is creating havoc in Australia's pristine marine ecosystems.

David Suzuki has made a marked impact for the better on the intellectual life of Australians, not least Tasmanians. He advanced our perception of the reality of *Homo sapiens*' unsustainable impact on Earth's life-sustaining biosphere. Though popular reaction has been suppressed by the insatiable demands and greenwashing of the corporate exploiters of our planet, his voice has advanced us nearer to the only solution to this disastrous epoch's trajectory: the end of mass denial and the taking of mass action.

David Suzuki is Earth's fortune.

With Bob Brown.

Peter Garrett

LEAD SINGER OF MIDNIGHT OIL,
ENVIRONMENTALIST, ACTIVIST,
FORMER POLITICIAN

David came into view in Australia in the late '80s and early '90s via his books, docos, and regular visits. He was instrumental in lifting awareness of the natural world at a crucial time of reawakening about the fate of our precious planet Earth. Prodigious scholarship, highly effective advocacy, and a campaigning instinct second to none, leavened by a great love of life, were among his formidable attributes and explain why he touched so many. I am privileged to consider him a "distant" friend, and he was a welcome visitor—completely at home with musos and fans—when Midnight Oil toured North America. It's a great innings and one can only wonder at, and be thankful for, the far-reaching contribution David has made.

Mike Rann

FORMER PREMIER OF SOUTH AUSTRALIA, GLOBAL CHAIR OF THE CLIMATE GROUP, PATRON OF THE WORLD SUSTAINABLE DEVELOPMENT FORUM

I met David thirty-five years ago when I was a young, new minister in the South Australian state government. I had been a big fan of his series *The Nature of Things* and his writing, so I invited him to come to Adelaide to launch and become the patron of the South Australian Youth Conservation Corps. It aimed to capture the enthusiasm of unemployed young people (many of whom had dropped out of school) for the environment in a series of conservation projects, to get them back into accredited training, further education, and careers. David accepted, and I was impressed with how he connected with and inspired these young people. I will always remember his kindness and walking with him through a beautiful conservation park north of Adelaide as he talked with my son David, then age six.

For many years, people came up to me in the street and told me about how their life trajectory and self-esteem had been changed for the better by their involvement in the Youth Conservation Corps. Many told me they were working in environment-related jobs.

For more than three decades, I have told people that David Suzuki has been and remains one of the most influential people in my life, including my political life and since then in my work with the Climate Group internationally with state and provincial governments on every continent. David has inspired so many of us not only through his books, television programs, and policy advocacy but also through his example.

When I became premier in 2002, I wanted South Australia to become a laboratory for reform and for our state to be a leader and an example not only in Australia but internationally. Inspired by David

Suzuki and helped by people such as Australian environmentalist and scientist Tim Flannery, we embarked on a big agenda of environmental and climate action. I kept David informed of our program, such as planting three million native trees in a series of urban forests to cool and beautify the Adelaide metropolitan area. Against much opposition, we established nineteen marine parks along our coast and doubled a vast area of land under wilderness protection. We solar-powered high-profile public buildings and schools and were the first Australian state to legislate for a feed-in tariff to reward householders for embracing solar power. Today, close to one in two homes have done so.

Perhaps most importantly, we were the first state in Australia to pass a climate change and carbon-reduction law that legislated a timetable and process to roll out renewable energy and reduce emissions. Despite considerable opposition, we exceeded all these targets. We were also the first jurisdiction in the world to establish interim targets for the percentage of our power generated by renewables in the long march toward net zero in 2050. From almost zero renewables, South Australia, with no hydro power in the driest state on the driest continent, will reach 100 percent renewable electricity by 2027. This is now seen as an exemplar for others around the world. In every speech, I mention that David Suzuki not only encouraged me but was my inspiration. Decades ago, David spoke to me about reducing plastic pollution, and I'm pleased we were one of the first jurisdictions in the world to ban non-reusable plastic bags.

I want to thank David for his encouragement, particularly the inspiration to keep going in the face of bitter opposition—including from climate deniers at the highest level in Australian politics who,

along with powerful media interests, are doing the bidding of a fossil fuel lobby whose destructive greed and dishonesty knows no limits.

David Suzuki has, like David Attenborough, been the persistent voice for nature and a compelling advocate for a sustainable life on Earth. He has encouraged countless numbers of us in politics and in the environmental movement internationally to keep going against formidable opponents. It's now our turn to urge David to "keep going." The world needs David Suzuki. I'm looking forward to celebrating his continued advocacy when he reaches one hundred. After all, it was David who encouraged us to listen to the wisdom of the elders. In the meantime, the forest named after him here in South Australia continues to flourish as recognition and a quiet symbol on the other side of our planet of David Suzuki's leadership.

Former Assembly of First Nations National Chief Ovide Mercredi and I were honoured at the University of Manitoba on October 24, 2014, during a Pipe Ceremony and Blanket Ceremony led by Elder Dave Courchene, for leadership in protecting the sacred land and advocating for Indigenous and treaty rights. Erica Daniels and Brandi Vezina, Turtle Lodge

10 | STARTING THE DAVID SUZUKI FOUNDATION

DURING THE 1980S, public awareness and concern about environmental issues rose greatly. In 1988, producer Anita Gordon asked me to host a CBC Radio series on environmental issues, five shows that were broadcast in a series called *It's a Matter of Survival.* I interviewed more than 150 scientists and experts from many countries and fields about environmental problems and how the world would look fifty years hence if we carried on with business as usual. The radio series conveyed the magnitude of the problem as well as the uplifting message that by acting now we could avoid the fate we were heading toward. More than sixteen thousand letters came in, most ending with the plea "What can I do?"

Tara said, "David, we've been warning people about the problems for years. This response shows we've reached a lot of the public, but now people feel helpless because they don't know what to do. You've got to go beyond the warnings and start talking about solutions."

Many of our friends had begun to challenge us to lead an initiative, perhaps to found a new organization. With their help, we invited about twenty "thinkers" to a weekend retreat to discuss whether we needed a

Back out in nature, filming in the Slocan Valley for *Force of Nature: The David Suzuki Movie.* Ari Gunnarsson

A group of thinkers got together on Pender Island in 1989 and gave birth to the David Suzuki Foundation one year later. Back row, L-R: Cliff Stainsby, Bill Green, Marilyn Van Bibber, Stan Persky, Bill Rees, Tara Cullis, Marina Lent, Russel Barsh, Tony Pearce, Irving Fox. Front row, L-R: me, Sue Ward, Doug Aberley, Severn Cullis-Suzuki, Sarika Cullis-Suzuki.

new solutions-oriented group. About a dozen people gathered for three days in November 1989 on Pender Island, one of the Gulf Islands in the Strait of Georgia. We agreed that most organizations had sprung up because of a crisis. But each crisis is merely a symptom or manifestation of a deeper, underlying root cause. An organization was needed to focus on root causes, so that steps could be taken to produce real change.

We agreed it should be science-based. We would use the best scientific information available and hire scientists to help write or edit the papers we wanted to produce. Further, we would learn how best to deliver this top-quality information to the public. When the group decided the name should be the David Suzuki Foundation, I objected. I was not in this endeavour to be remembered in perpetuity nor forever saddled with responsibility of ensuring the organization lived up to its promises and programs because of my name.

The counter-argument was twofold. First, my profile in Canada had been built up over many years of working in science and the media and of speaking out, so my name would immediately tell people what the foundation stood for. Second, it might be possible to translate the reputation my work had created into fuel for the new organization. By the end of the weekend, the Pender Island retreat had created a new organization.

Miles Richardson, then president of the Haida Nation, was one of the first three board members, along with Tara and me. One of the strengths of the foundation from the beginning has been ongoing Indigenous input.

By September 1990, Tara had established our legal status as a charitable organization. Shortly after that, she found space for an office that was formally opened on January 1, 1991. Over the years, thousands of people had written to *The Nature of Things With David Suzuki*. Many letters were addressed to me personally. I felt that if someone had taken the time to write a letter, they deserved a response. All those people I wrote back to and the sixteen thousand who responded to *It's a Matter of Survival* made for a wonderful list of people we could approach for support. We drafted a letter reminding people they'd once written to me and asking for their support to find solutions to the ecological degradation of the planet.

With the help of many volunteers, in November 1990 Tara sent out some 25,000 letters. And then, just before Christmas, cheques and cash began to come in, first as a trickle and then as a flood of full mailbags. Within months, people were writing to ask what we had accomplished with their money.

Tara and I had already been investing our own money to support Barbara Zimmerman, who was working in Brazil with Paiakan and the Kaiapo of A'Ukre to establish a research station in the Xingu River watershed. We transferred this project to the foundation. I had also been introduced to the Ainu of Japan, Indigenous people who had held on to their culture through 1,500 years of Japanese occupation. Now they were close to losing their language and their last sacred river, the Saru, which was to be dammed to provide energy for industrial development. We also turned

I am feted by Ainu People in Hokkaido while wearing a blanket given to me, along with name, Na̲nwak̲awe, by Ernest Alfred of the 'Na̲mg̲is First Nation.

over this project to the foundation. Unfortunately, the dam was built a few years later.

Gradually we raised enough money to hire staff and an executive director. We received several applications and winnowed them to a shortlist that included Jim Fulton, the Canadian member of Parliament who had tipped me off about the struggle over logging in Windy Bay in Haida Gwaii. It became clear that Jim's track record as a committed environmentalist, his experience as a politician, the high esteem in which he was held by First Nations and communities, his irresistible personality, and his exuberant energy made him the best choice. We were thrilled when Jim accepted our offer.

By the time we hired him, we had already begun to acquire the financial support that enabled us to move to a new office on Fourth Avenue in Vancouver, in the heart of the Kitsilano neighbourhood. We sponsored a conference in May 1995 and invited people who have studied and helped influence social change to share their insights; those talks were published as "Tools for Change," a document that has been infused in the way we do our work.

As we began to scope out our first project on fisheries, it became the model for later work. The five species of Pacific salmon—sockeye, pink, chum, Chinook (or spring), and coho—lie at the heart of First Nations cultures on the West Coast, nourishing them physically and spiritually. Carl Walters, a world-renowned fisheries authority at the University of British Columbia, accepted our invitation to write a scientifically based evaluation of the state of Pacific salmon. The report, "Fish on the Line,"

Jim Fulton (1950–2008), former member of Parliament and first executive director of the David Suzuki Foundation, at play.

concluded that salmon runs were in trouble along the B.C. coast. Since then, the foundation has funded numerous fisheries projects and reports, including original research by Tom Reimchen of the University of Victoria about the interconnections between salmon, the Pacific rainforest, and the life salmon sustain.

Salmon are at the centre of one of our most gratifying projects, revitalizing the run to Musqueam Creek in Vancouver. In 1996, the David Suzuki Foundation was approached by the Musqueam People to help rehabilitate the creek. Willard Sparrow, grandson of the famous Chief Edward Sparrow Jr., had become concerned. Nicholas Scapiletti was working for the foundation and hit it off with Willard as the two of them campaigned to raise money to clean up and protect "Vancouver's last salmon creek" and to educate people in the neighbourhood. Musqueam Creek is now on its way back to health.

Now the foundation has matured. We have earned a presence in the media, influence within the political and industrial community, and credibility with the public.

Tara delivers a speech as president of the David Suzuki Foundation.

Stephen Bronfman

EXECUTIVE CHAIRMAN OF CLARIDGE, DAVID SUZUKI FOUNDATION BOARD CO-CHAIR

Meredith Webster held a cocktail party for David Suzuki and Mohawk Chief Joe Norton at her house in the fall of 1989. I was twenty-five. My mother Barbara (knowing I was a big fan of David's and of the natural world in general) asked me if I wanted to join her and meet David and Chief Norton.

Yes, of course I did.

That first meeting led to a thirty-five-year relationship between me, David, Tara Cullis, the Suzuki family, and the David Suzuki Foundation.

I volunteered from the beginning to do work for the foundation. They sent me to Tasmania to work with Bob Brown of the Green Party on lobbying the Australian government to dismantle a dam to restore Lake Pedder. That dam flooded large tracts of open land in central Tasmania. I was honoured to represent the foundation.

I've been working with the David Suzuki Foundation ever since and have been a dedicated board member for over thirty years. David, Tara, Severn, and Sarika have been (aside from my parents) my greatest mentors in my life's journey.

I don't know of a family more dedicated to our natural world. All are brilliant, all hold doctorates, and all have the gift of teaching science to the average person.

David Suzuki is an icon in Canada, an icon in the world of climate and environmental science.

It is hard for me to convey how meaningful this relationship has been throughout my adult life. It's such a great honour and privilege for me to call David a dear friend and brother in the fight for what is so needed for our Mother Earth.

Elois Yaxley

EXECUTIVE ASSISTANT TO DAVID SUZUKI, 2002–12

One of the highlights of my life was working as David's executive assistant from 2002 to 2012. It was an exciting decade trying to keep David's incredibly busy work life in order. Every morning felt like a new adventure. Who knew who might call to speak or meet with David? It could be random calls from the prime minister's office, Wesley Snipes wanting to do lunch, Bruce Cockburn asking David to help with a fundraising event for the Unitarian Universalist Service Committee, or the Cannes Film Festival requesting David's presence for a screening of the film *Force of Nature: The David Suzuki Movie*.

Working as David's EA was exhausting and demanding at times as I tried to keep many balls in the air. Any one of David's commitments would be a full-time job for most people. David hosted CBC's *The Nature of Things* for forty-four years and was consequently a fan favourite for all CBC events that needed a high-profile face in attendance. David rarely said no to requests for his time from CBC staff. He had a prolific writing career, to date writing more than fifty books, so there were often commitments with his publishing company to fit into his schedule, followed by cross-Canada book tours. As the chair of the David Suzuki Foundation, David was also in demand from staff and board members, and I was the gatekeeper. In addition to foundation commitments, David was a popular guest speaker at conferences and on average committed to at least thirty speaking engagements a year. Then there was his mail, email, and telephone requests from fans and a few haters. David thought it was important to respond to everyone.

When I faced personal challenges during the decade I worked for David, he never hesitated to support me when I asked for time off or an adjusted work schedule. If I could keep everything running smoothly, David was happy.

David's humanity, compassion, and love shone through in the fall of 2021, nine years after I left his employ. My fifty-year-old daughter, Gro Kathleen Averill, was diagnosed with terminal cancer. The moment David heard about her diagnosis, he called. Although he was extremely busy, David took the time to offer words of caring and expressions of love to my entire family. On the last morning of my daughter's life, when she chose medically assisted death, David called to say we were in his thoughts, and he was sending me a hug and asked me to give my daughter a hug.

I remember hearing David say in numerous interviews that the most important thing in life is love. In his book *The Sacred Balance*, he devoted a chapter to "the law of love": Love is what makes us human. David's words of comfort helped me bring love to my daughter and to surround her with love and support during the last

Foundation staff and I bid farewell to long-time executive assistant Elois Yaxley (standing far left).

two hours of what she called her "Celebration of Life." David's phone call gave me strength as I read Kahlil Gibran's poem on love, as Gro had requested.

David is concerned about the well-being of the planet and has spent his life working tirelessly to educate folks on the importance of clean air to breathe, clean water to drink, and a planet with healthy soil that can sustain life. David Suzuki is a genuinely compassionate human being who is not afraid to express his love for others, including his executive assistant.

A few weeks after my daughter died, David and Tara stopped by my home to give me a hug and have a cup of tea. "How are you doing?" David asked. Simple words, but extremely meaningful and profound to a person going through the loss of a dearly beloved. David and Tara gave our family a five-hundred-piece puzzle of gorgeous butterflies. It was an opportunity for our family to restore some order to a horrific time. Putting puzzle pieces together became an active meditation for our family that provided a glimmer of hope for the days that lay ahead. The result was a beautiful picture of many butterflies that reminded us of Gro.

Thank you, David, for your caring and compassion that helped me on my healing journey.

David Boyd

FORMER UN SPECIAL RAPPORTEUR ON THE RIGHT TO A HEALTHY ENVIRONMENT; ASSOCIATE PROFESSOR OF LAW, POLICY, AND SUSTAINABILITY, INSTITUTE FOR RESOURCES, ENVIRONMENT AND SUSTAINABILITY, SCHOOL OF PUBLIC POLICY AND GLOBAL AFFAIRS, UNIVERSITY OF BRITISH COLUMBIA

David Suzuki is one of my heroes, and he's one of the main reasons I chose to become an environmental lawyer more than thirty years ago. It was 1992. I had a law degree gathering dust while I fiddled around figuring out how to become a writer. Then I read the bestseller *It's a Matter of Survival,* co-authored by David and Anita Gordon. It was one of those light bulb moments, when I realized I could and should harness my legal education to help protect the planet. Within months, I was volunteering for the Sierra Legal Defence Fund (now Ecojustice), where I subsequently articled and fell in love with being an environmental lawyer.

Of course, the David Suzuki Foundation turned out to be a client, and I enjoyed the opportunity to get to know David and appreciate his vastly underrated sense of humour. At a mid-1990s winter party in a bowling alley, David was wearing a name tag that said "Bill." A young volunteer with the David Suzuki Foundation, who perhaps had imbibed one too many Moosehead beers, pulled me aside. "Excuse me," he whispered, "but that guy named Bill really looks like David Suzuki." It's his cousin, I explained. Years later, wandering with David down a wind-blasted street in downtown Winnipeg, two young men (also apparently prodigious consumers of Moosehead) approached us and said, "Hey man, you really look like David Suzuki." David tactfully said, "Yeah, it's funny, a lot of people think that. But he's much shorter than I am."

I can also blame David for the worst year of my professional career. After I'd written my first book, a doorstopper called *Unnatural*

Law: Rethinking Canadian Environmental Law and Policy, I asked David to provide one of his pithy blurbs for the back cover. David kindly agreed, but after I sent him a draft, he bluntly told me I needed a shorter version that politicians could read. I agreed, and the foundation generously published *Sustainability Within a Generation: A New Vision for Canada*. Unbeknown to me, David handed a copy of the report to then Prime Minister Paul Martin at a meeting in Ottawa. Surprisingly, Martin read the report and called David to tell him he agreed with it and wondered whether the author might come to Ottawa to work for him. Knowing I was a dedicated West Coaster, David told the PM to forget about it. News of this conversation didn't reach me until several months later through Jim Fulton, the David Suzuki Foundation's charismatic leader. My jaw dropped. Of course I'd work for the freaking PM! And so I spent an *annus horribilis* in Ottawa, learning the dark secrets of how the federal government worked, until NDP leader Jack Layton pulled the plug on the minority Martin government and paved the way for Prime Minister Stephen Harper.

Finally, David helped bend my career in a new direction by being super stoked about my doctoral research on the human right to a healthy environment. The foundation ran the Blue Dot campaign, harnessing a national network of volunteers to convince almost two hundred municipalities to adopt resolutions supporting this fundamental right. That campaign paved the way for Canada to vote in favour of a historic United Nations resolution recognizing the right in 2022, and for the addition of every Canadian's right to a healthy

environment to the Canadian Environmental Protection Act in 2023. The feds put the words into action in 2024, creating a new rule on benzene emissions from the petrochemical industry to protect the right to a healthy environment of the Aamjiwnaang First Nation near Sarnia, Ontario. A major polluter chose to close its factory forever instead of installing a cleaner production process, with benefits for both human and ecosystem health.

David Suzuki's fingerprints are all over my career, and I'm immensely grateful for his support and friendship. He's an inspiring visionary who, despite his modesty, has contributed to countless environmental victories around this beautiful planet through his own actions and the actions of millions of people like me whom he inspired to take environmental action.

A cold dip to celebrate the end of dumping raw sewage into Halifax Harbour, during the 2014 Blue Dot Tour.

DAVID SUZUKI
THE SACRED BALANCE
DAVID
SUZUKI

11 | UP AND RUNNING

IN ITS EARLY YEARS, the David Suzuki Foundation picked up experience in organizing news conferences and writing news releases, articles, opinion articles, and other documents, and the day came when the communications group, headed by David Hocking, established a website.

James (Jim) Hoggan, president of the largest communications and public relations company in western Canada, offered his expertise on a voluntary basis. He helped us develop the most effective ways to get our message out, and ever since he joined the board, he has devoted countless hours to our communications efforts.

It was hard work to get media attention until foundation employee Catherine Fitzpatrick had the brilliant idea of looking at the medical implications of burning fossil fuels. She concentrated on the direct, day-to-day, physical effects of air pollution on people. Using government data, the doctors and scientists Catherine commissioned produced a report entitled "Taking Our Breath Away." It found that air pollution, much of it from burning fossil fuels, was prematurely killing sixteen thousand Canadians a year. "Taking Our Breath Away" was the first foundation report to be translated into Canada's other official language, French, as "À couper le souffle."

Signing books after my keynote speech at a conference in Toronto, June 2018. Shez Mehra

After the Earth Summit in Rio in 1992, an intergovernmental negotiating committee was established to meet and work out a framework within which the climate convention could be assessed. In 1995, the Conference of the Parties (COP) to the United Nations Framework on Climate Change was established to meet annually in a different country. The first COP meeting in North America was held in Montreal from the end of November to early December 2005. Thousands of delegates, non-governmental organization participants, and media attended, and the foundation was a prominent participant.

Many of us in the David Suzuki Foundation cut our teeth on battles over the future of British Columbia's forests. Richard Schwindt and Terry Heaps, economists at Simon Fraser University, agreed to do an analysis of the forest industry, and we published it in 1996 as "Chopping Up the Money Tree." In 1990, foundation staff member Ronnie Drever wrote a report published as "A Cut Above," which outlined nine basic principles of

The David Suzuki Foundation board in the early days. Back row, L–R: Wade Davis, Mike Robinson, Peter Steele, Stephen Bronfman, Jim Fulton. Front row, L–R: Severn Cullis-Suzuki, me, Tara Cullis, Ray Anderson, Jim Hoggan. (Not pictured: Stephanie Green and Miles Richardson.)

what has since come to be called ecosystem-based management. We also published a report on culturally modified trees (CMTs)—for millennia, B.C. coastal First Nations used parts of trees like bark and sap without cutting the trees down.

In addition to the report "The Cultural and Archaeological Significance of Culturally Modified Trees," we initiated an archaeological training program that enrolled dozens of representatives from eleven coastal villages to become CMT technicians.

We wanted the findings from our reports to reach a wider audience, and in 1994 the foundation met with Greystone Books, then a division of Douglas & McIntyre. The foundation and Greystone would co-publish books meant for a wide audience, and we hoped the books would find a readership big enough to make them relatively self-sufficient. Government restrictions on charitable activities later necessitated transferring that relationship to the David Suzuki Institute, which is not a registered charity.

As the foundation grew, it seemed to me we needed a philosophical treatise that would define the perspective, assumptions, and values that underlie our activities. *The Sacred Balance: Rediscovering Our Place in Nature*, published in 1997, expressed what I had learned since my interview with Guujaaw that brought the fundamental issues into sharp focus for me. The book became a number one bestseller in Canada and Australia and continues to sell. It was updated in 2022 for a twenty-fifth anniversary edition.

The foundation and Greystone's first joint book launch, for *Dead Reckoning* by Terry Glavin, at Cannery Restaurant, 1996. L-R: me, Jim Fulton, Suzanne Hawkes, Rob Sanders. Janice Williams

The foundation's cornerstone is its relationship with Indigenous Peoples and communities. In 1998, an opportunity arose to work on coastal issues in B.C. Miles Richardson, one of the foundation's founders, urged us to look at forming relationships with First Nations communities. In the winter of 1997–98, Jim Fulton asked Tara to step into a staff position. He believed the best way to protect the forests and fish was to work with First Nations to help realize their sovereignty over the territory. This approach was met with skepticism or even downright hostility from some who believed that if given sovereignty, Indigenous people would be just as destructive. We believed that thousands of years of living in place shaped their relationship with their territory. Tara began to travel alone into each community to meet the leaders, Elders, and families in the villages.

Long before these first forays, Tara (and our family) had already established deep friendships in two of these villages, Skidegate and Bella Bella, as well as Alert Bay and others to the south. We had been adopted by two families and had long felt a responsibility to contribute to their villages. Although conservation was a serious issue to all the villages, jobs—community economic development (CED)—were the priority. We accepted this challenge and set about turning ourselves into a CED organization, opening a foundation office in Prince Rupert.

In British Columbia, most First Nations are represented in what's called the First Nations Summit; delegates meet in Vancouver regularly to discuss issues of mutual interest. In the fall of 1999, we invited leaders from the communities Tara had been visiting to meet us and each other to discuss some forestry information.

At the meeting, Art Sterritt from the Gitga'at community of Hartley Bay and Gerald Amos of Kitamaat village suggested the foundation call a conference of all communities in the central and north coast and Haida Gwaii. We invited the eleven communities, plus Nemiah, to a two-day meeting at the Musqueam reserve in Vancouver in March 2000 and raised money to pay all expenses.

To commemorate the millennium, the meeting was called Turning Point. The foundation attended in a supportive role, providing funding, organization, research, and contacts. We made it clear that while we believed the land belonged to the First Nations communities and supported their struggle to have that ownership recognized by government, we wanted the forest and marine ecosystems in which they dwell to remain healthy and productive in perpetuity.

On April 4, 2001, B.C. Premier Ujjal Dosanjh signed two documents, one of which set in motion negotiations with the provincial and federal governments and the Turning Point communities on what was termed a government-to-government-to-government basis.

In March 2000, the foundation helped host a conference of First Nations from the central and north coast and Haida Gwaii in B.C., leading to the signing of the Turning Point agreement.

As the Turning Point organization grew in strength and independence, the foundation's role diminished, and in September 2003, in a formal celebration in Skidegate village on Haida Gwaii, we were thanked and farewells were made. The foundation received a drum, as a symbol of the heartbeat of the people, and we gave each community a gift of fossilized cedar leaves, a symbol of tenacity and survival that I had presciently bought in Drumheller years before. Turning Point now lives on as the Coastal First Nations–Great Bear Initiative.

Preparing the twins for the Salmon Dance, Alert Bay, 2024. Back row, L-R: Tidi Nelson, Ryo Suzuki Rhodes, Christopher Rhodes, Ernest Alfred, me, Tara. Front row, L-R: Sarika Cullis-Suzuki, twins Kaoru and Nakina.

Andrea Seale

CHIEF EXECUTIVE OFFICER OF THE CANADIAN CANCER SOCIETY, FORMER CEO OF THE DAVID SUZUKI FOUNDATION

Working alongside David Suzuki for eight years at the David Suzuki Foundation was one of the most transformative experiences of my life. His influence on me went far beyond the workplace—he helped shape my world view, deepened my appreciation for the power of community, and instilled in me a profound love and respect for the interconnectedness of all life.

David inspired through his words and his communications and through his presence. He came to staff meetings almost every week. He came to challenge us, to ignite our passion—and he always did! He would tell us stories, offer sharp critiques of politics, challenge us to do more, and—unforgettably—one day he showed us how to twerk! David brought his drive, a little bit of humour, and total authenticity to every situation.

He pushed us to think bigger about what's possible—how to create a bigger movement that would welcome all. Working on the Blue Dot Tour and the Radically Canadian campaign were two highlights for me. I appreciated that in the face of so much destruction of the planet, he gave hope by reminding us that nature could rebound—if only we let her. His passion for the planet and for social justice was contagious. He reminded us constantly that environmentalism isn't just about saving trees or protecting species; it's about people, systems of power, interconnected relationships, and a future where we live in greater balance with all other species.

David's legacy is immense. He has given so much to Canada, to the world, and to every person lucky enough to work with him.

Nancy Flight

EDITOR EMERITA,
GREYSTONE BOOKS

In 1980, when I was an editor at an academic publishing company in San Francisco, I was charged with editing the second edition of a highly successful genetics textbook written by two professors at UBC—David Suzuki and Tony Griffiths. I had no idea what a massive effect this would have on my life.

When I flew to Vancouver to meet with the authors, David invited me to dinner along with Tara and, for some strange reason, Tara's brother, Pieter Cullis. Pieter asked me to go out with him the next night, and within six months we were married, and I somehow found myself living in Vancouver.

David was as stunned as I was by this result, but he accepted this turn of events with grace (even though, as I later discovered, he had been referring to me as the "Dragon Lady") and even suggested to his Canadian publishers that I edit a scientific book he was writing for them, having somehow gotten the idea that I knew something about science. Later, when David's publisher went out of business, David decided to sign on with Greystone Books, where, unbeknown to him, I had recently accepted a position as editor. He just could not get rid of me.

Working with David on such books as *The Sacred Balance, David Suzuki: The Autobiography, You Are the Earth*, and many others has been a highlight of my fifty years of editing, and I think he only yelled at me once (at least that is all I have chosen to remember). David is

remarkable in so many ways, but I think what impresses me most is his interest in other people. He is always curious about what other people are doing and what they think—and, of course, how their love life is going. I am grateful to David for so much—for introducing me to Pieter, for somehow putting up with me both as his sister-in-law and as his editor, for all he has done for the planet, and, of course, for his love of gossip. Here's to many more years of fighting for our planet, writing books, and passing along juicy tidbits.

Speaking at a book launch for *The Sacred Balance* in 1997. Janice Williams

James Hoggan

PRESIDENT OF HOGGAN AND ASSOCIATES, DAVID SUZUKI FOUNDATION BOARD MEMBER

I've learned a lot about doing the right thing from David Suzuki over the thirty-five years I've known him. One lesson began late in spring 2007 on a beach in Kauai where I was nervously writing a speech. I had been invited to give the Diana and Charles Tisdall Lecture in Communication at the national conference of the Canadian Public Relations Society (CPRS). It was an important speech on ethics in public relations, with a large crowd.

Back then, I owned a high-profile PR agency. I was chair of the David Suzuki Foundation and of Al Gore's Climate Project Canada. I had started the online news site DeSmog to clear up the PR disinformation about climate science. The title of my speech was "You Can't Spin Mother Nature." I planned to say that public relations people shouldn't help their clients mislead the public about climate change. The conference was in Edmonton, Alberta, near Canada's largest source of greenhouse gas emissions, the oil sands. I was nervous and about to get more nervous.

Kevin Grandia, who managed DeSmog, called. He apologized for interrupting my vacation but said the CPRS board of directors wanted to talk to me. A group of public relations people had complained that I was giving the keynote speech and demanded that the CPRS cancel my talk. They had started a campaign with a nasty website accusing me of all sorts of made-up things.

I tend to get nervous about speeches anyway, but the thought of people walking out of my speech en masse made me more so. I decided to ask David Suzuki for advice on giving a speech to an unsupportive crowd. After I told him what was happening, I asked,

"David, you've given speeches where some of the crowd was not happy with what you had to say. Any advice for me?" He replied, "Have they threatened to beat you up?" I said no. David said, "Well, that's what they used to do to me. Give the speech, don't hold back, tell them exactly what you think. I bet you get a standing ovation."

I spoke to the CPRS board and told them I had no intention of changing my speech to satisfy people who want to spread misinformation about climate change.

In the end, David's advice worked. I got a standing ovation. This was a big lesson for me on the need to speak up and set aside the fear of repercussions. Propaganda can work through stealth. Propagandists do not have to silence us if we silence ourselves out of fear. Self-silencing out of fear threatens freedom of speech and social change. The authoritarian propaganda project polluting the commons today is counting on our silence. Thanks for the advice, David.

Jim Hoggan has devoted countless hours to the foundation, as a communications adviser, board member and chair, and friend. Samantha J. Walker

ARUBA

12 | RIO AND THE EARTH SUMMIT

WHEN WE RETURNED from our trip to the Amazon in 1989, Severn was so upset after seeing the rainforest under assault by farmers and gold miners that she started a club made up of her grade five friends who shared her concern about forests. They called their club the Environmental Children's Organization (ECO) and began to give talks at schools about the importance of protecting forests.

In 1991, Severn heard about the Earth Summit to be held in Rio de Janeiro in June 1992 and asked whether Tara and I were going. I answered that we weren't and asked why she was curious. "Because I think all those grown-ups are going to meet and make decisions, and they're not even going to think about us kids," she answered. "I think ECO should go to remind them to think about us."

Without even reflecting on Severn's idea, I rejected it. Tara and I thought kids would be overlooked. Besides, it would be hot, crowded, and polluted.

That summer, we had a visit from Doug Tompkins, an American who started the clothing company Esprit with his wife, Susie Russell. When the marriage broke up, Susie bought him out and left him with a considerable chunk of money. He had decided to commit his

Sev and a woolly monkey at the Ariau Towers in the Amazon.

money to environmental causes, especially buying large tracts of wilderness to protect.

Somehow he had heard about Tara and me and the newly formed David Suzuki Foundation, so he flew to British Columbia with deep ecologist Bill Devall and visited us at our cottage on Quadra Island. Severn told Doug about her idea of taking ECO to Rio the following year. He told her, "That's a good idea. Write to me about it." She did, and a couple of months later, she said, "Hey, Dad, look," and held up a cheque for US$1,000 from Doug Tompkins. Tompkins later ended up focusing on the great cedar forests of Chile, where he purchased hundreds of thousands of acres and moved there to help local communities develop a sustainable relationship with the forest. Tragically, he died in 2015 from hypothermia after his kayak overturned in a storm.

For the first time, I realized Sev could be onto something. Tara and I decided it might be worth going to the Earth Summit. So we told Sev we would support her by matching every dollar ECO could raise. That meant she already had $2,000.

The ECO gang in Rio. L-R: Michelle Quigg, Severn Cullis-Suzuki, Raffi, Sarika Cullis-Suzuki, Morgan Geisler, Vanessa Suttie.

Severn and Sarika and the other ECO girls plunged into projects to raise money. In all, they raised more than $13,000, which Tara and I had to match—enough money to send five ECO members (including, of course, Severn and Sarika) and three parents (Tara, me, and Patricia Hernandez, the mother of one of the other girls) to Rio.

With the ECO gang in Brazil.

For me, the important thing was to establish the real bottom line: that we are biological beings, completely dependent for our good health and survival on the health of the biosphere. At the David Suzuki Foundation, we felt one contribution we could make at Rio would be such a statement or vision, so I began to draft a declaration. As I began to work on it with Tara, she suggested it should be like the American Declaration of Independence, a powerful document that would touch people's hearts. "How about calling it the Declaration of Interdependence?" she suggested.

Tara and I went back and forth with our efforts and then recruited children's songwriter Raffi, our Haida friend Guujaaw, and the Canadian ethnobotanist Wade Davis to contribute. At one point I kept writing the cumbersome sentence "We are made up of the air we breathe, we are inflated by water and created by earth through the food we consume." As I struggled with the lines, I suddenly cut through it all and wrote, "We are the earth."

That was the first time I really understood the depth of what I had learned from Guujaaw and other Indigenous people. I knew we incorporate air, water, and earth into our bodies, but simply declaring that's what we are cut through all the boundaries. I understood that there is no line or border that separates us from the rest of the world.

I believe the final Declaration of Interdependence (davidsuzuki.org/about/declaration-of-interdependence/) is a powerful, moving document that sets forth the principles that should underlie all of our activities. We had the declaration translated into many languages and took copies to give away in Rio, one of the first tangible products of the David Suzuki Foundation.

At one of the Earth Parliament events, I gave a short talk and then yielded the stage so that Severn could make a speech. In the audience was James Grant, the American head of UNICEF, the United Nations Children's Fund, and he was so moved by Sev's remarks that he asked for a copy of them. We learned later that he had run into Canadian Maurice

Strong, one of the driving forces behind the establishment of the United Nations Environment Programme (UNEP) and told Maurice he should give Severn an opportunity to speak.

The next morning, we received a call from Strong's office inviting Severn to speak at Rio Centro in a session for children. We didn't have much time. Sev wrote out her speech on a piece of paper, adding words and phrases in the margins as all of us offered our critique. At one point, I said, "Sev, here's what you should say," to which she retorted, "Dad, I know what I want to say. I need you to show me how to say it." She was right. The power of her words, like Greta Thunberg's thirty years later, was that they came from a child's perspective unclouded with a hidden agenda.

We entered the conference room. It looked almost empty. Sev was last on the list. The other girls made their presentations well, pleading for better care of resources, wildlife, water, and the poor. Finally, it was Sev's turn. She was twelve years old and had not had time to prepare thoroughly, and I was scared stiff—but I hadn't given her enough credit.

Severn speaks to the plenary session of the Earth Summit in Rio de Janeiro.

I am told that when Sev began to speak, people in the halls stopped what they were doing and gathered around the television sets to listen. When she left the stage to come to us near the middle of the auditorium, her first words to Tara were "Mommy, could you hear my heart beating?" As she sat down between us, a member of the American delegation rushed over to shake her hand and congratulate her. "That was the best speech anyone has given here," then U.S. senator Al Gore told her.

It was all heady stuff for a twelve-year-old, and I began to worry about what this would do to her sense of herself. I stopped worrying the next year when she was invited to appear on *The Joan Rivers Show* in New York. "Dad," Sev told me, "I hope it's okay, but I'm not going to do this. I have to study, and I want to make the basketball team." She had her priorities right.

You can watch Sev's speech at youtu.be/F_O1Au8vZLA.

Maude Barlow

AUTHOR, ACTIVIST

David has touched so many lives. On a lovely June evening in 2018, he inspired my then sixteen-year-old granddaughter Eleanor to choose a study and career path dedicated to protecting Earth. David and I, along with Avi Lewis, were on a panel on the climate crisis in an old church in Ottawa. The church was packed, especially with young people. David was on fire that night—a preacher for the biosphere. He laid out the facts of the environmental Armageddon we face if we don't change course. But he also laid out the steps we have to take, individually and collectively, to heal nature. Eleanor was in awe.

After the panel, I brought her backstage to meet David. She was nervous! What would she say to the great David Suzuki? But of course, David was so warm with her, asking her about her thoughts and plans and listening to her as if she were the most important person in the world. As we walked home that evening, the air fragrant with apple blossoms and lilacs, the newly leafed trees silvered by a new moon, Eleanor and I talked about her future.

I am proud to say she recently graduated with her master's in environmental studies and agriculture from the University of Guelph. No greater legacy than this is possible in my view.

We have been on many a stage together, and David always inspires me. I believe I have quoted him in almost every book I have written. I especially love his thoughts that it is necessary to allow for grief as we fight for Earth, but it is also necessary that we use that grief to overcome fear and spur us to action.

Wade Davis

AUTHOR, PROFESSOR EMERITUS OF ANTHROPOLOGY AT THE UNIVERSITY OF BRITISH COLUMBIA

When I was young, just getting people to stop throwing garbage out of car windows was considered a great environmental victory. No one spoke of the biosphere or biodiversity. Today these words are part of the vernacular of schoolchildren. In 1971, the year I entered college, Canada became just the second nation in the world to establish an environment ministry.

Those who spoke for Earth: Rachel Carson with her elegiac *Silent Spring* (1962); Tom Lovejoy and E. O. Wilson who together coined the term *biodiversity* (1980); Lynn Margulis and James Lovelock who celebrated the planet as Gaia, a living being (1972); the poet Gary Snyder, prophet of deep ecology, together with the wide-eyed rainbow warriors who first sailed under the banner of Greenpeace in 1971. All these were lone voices in the wild. To stand as they did, to say what they said, to challenge and defy scientific orthodoxies while calling into question an economic paradigm that laid waste to nature implied immense personal and professional risks, a level of moral courage difficult to imagine today.

Among the most influential and inspiring of these early voices was that of David Suzuki. As the global environmental movement took flight, David was an internationally renowned geneticist, precociously ensconced as a tenured professor at the University of British Columbia and yet driven by the certain conviction that the lessons and insights of science were too important to be sequestered within the isolation of the academic community. With radio's *Quirks & Quarks* (1975) and later TV's *The Nature of Things* (1979), his passion and curiosity opened Canada's eyes to both the wonder of science and the perfect beauty of the natural world. As a storyteller, he imbued nature with a sense of the sacred, knowing that to do so was not contrary to science but, rather, an acknowledgement of the complexity and wonder of ecological and biological systems that science has illuminated.

Many years ago, when I was starting out as a writer, David endorsed one of my books with a line that my publishers have been using ever since. It was a beautiful gesture, typical of his immense generosity of spirit, his willingness always to lend his name and support to all those who share his passion for the well-being of nature.

He may have intended it for me, but it could have just as well, and surely more deservedly, been written for him, for David is indeed "a rare combination of scientist, scholar, poet, and passionate defender of all of life's diversity."

I have known David as a friend and a mentor for half my life. Just to walk in his shadow is to aspire to greatness, to want always to do more, knowing that the real work of conservation is a pilgrim's path that has no destination, no goal save that of an utterly new dream of Earth.

TREJK
ATET

13 | KYOTO, PARIS, AND CLIMATE CHANGE

IN THE AUTUMN OF 1988, climate experts from around the world gathered in Toronto for a major conference on the atmosphere, The Changing Atmosphere: Implications for Global Security. Three hundred delegates, including representatives of more than forty governments, were welcomed by newly re-elected prime minister Brian Mulroney. Norwegian prime minister Gro Harlem Brundtland gave a keynote address, and Stephen Lewis chaired the sessions. Climatologist James Hansen, who had just told U.S. Congress that he was "99 percent sure" the heat spell America was going through was caused by human activity, also spoke. The delegates were warned that the threat of global warming was real and called for a 20 percent reduction in greenhouse gas emissions in fifteen years.

That year, the World Meteorological Organization and the United Nations Environment Programme established the Intergovernmental Panel on Climate Change (IPCC), made up of hundreds of climatologists from many countries, to monitor the state of global climate. Sadly, hindsight reveals that had governments responded and met that challenge beginning in 1988, the air today would be cleaner, people healthier, and fossil fuels more plentiful, and we would be saving hundreds of billions

Talking to Greta Thunberg at the 2019 climate march in Montreal.

of dollars and be well along the path to achieving an emissions level that could be absorbed by the biosphere.

U.S. president George H. W. Bush, an oilman, had refused to attend the 1992 Earth Summit in Rio de Janeiro unless the greenhouse gas emissions target was reduced. As a result, the target was shifted. Countries agreed to stabilize greenhouse gas emissions at the 1990 levels by 2000. Most countries, including Canada, merely called for "voluntary compliance" with the watered-down target. In the meantime, the fossil fuel industry launched an aggressive campaign to discredit the very idea that human activity was influencing climate, and the use of fossil fuels and thus greenhouse gas emissions continued to rise.

News conference using pop cans to represent greenhouse gas emission levels. L-R: me, Steven Guilbeault (Greenpeace Canada, now member of Parliament), Louise Comeau (Sierra Club Canada).

In 1995, to film for *The Nature of Things*, I attended a conference on climate organized by the Intergovernmental Panel on Climate Change in Geneva. Hundreds of IPCC climatologists from more than seventy nations had painstakingly assessed thousands of scientific papers on weather and climate, and they concluded in 1990 in their first major assessment that global climate was warming and that the change was not part of a natural cycle. In 1995, the IPCC's second assessment concluded that "the balance of evidence suggests a discernible human influence on global climate." Subsequent assessments have been increasingly stronger. The sixth assessment, from 2022, concluded that human influence on the climate is unequivocal, that global average temperatures have reached their highest levels in human history, and that immediate and deep emissions reductions are required across all sectors to limit heating to 1.5°C.

By 1997, global concern about climate change had grown enough to warrant a gathering of delegates from most countries in the world at Kyoto, Japan. I found myself reluctantly attending this conference along with David Suzuki Foundation staff. I say reluctantly because, at these massive international affairs, much of the decision-making goes on behind closed doors while groups such as ours merely buzz around like annoying gnats.

At the meetings, leading scientists talked about the latest evidence for climate change, environmental groups called for serious cuts in emissions, and government delegates wrestled with lobbyists working to sabotage the process by driving it off the rails. Alberta's delegates were hostile to

Speaking on a panel at the 2022 COP15 biodiversity summit in Montreal. Leo Van Doormaal

us, but Steven Guilbeault (then with Greenpeace and now a member of Parliament) came by to support us, as did the Quebec attendees. The insurance industry was the one large group in the business community that took climate change seriously. Their actuarial data were dramatic—claims for climate-related damage like fires, floods, droughts, and storms were rising exponentially, as were the number of insurance companies going out of business.

The European Union was concerned about climate change and wanted serious cuts in the range of 15 percent below 1990 emission levels. Aligned against them were the JUSCANZ countries (Japan, United States, Canada, Australia, and New Zealand), which formed a bloc working to water down the target.

LEFT: Former U.S. vice-president Al Gore has been a long-time champion for the environment and good friend. RIGHT: With NHL player Andrew Ference. The foundation and I have found innovative ways to champion climate action, such as getting NHL players to sign a pledge.

Kyoto signalled the recognition that the atmosphere is finite, that human activity has saturated it with emissions from fossil-fuel-burning vehicles and industries, and that we are adding more carbon dioxide and other greenhouse gases than the biosphere can handle. For the first time, governments and industries had to acknowledge that there can't be endless growth.

It began to look as if the proceedings would fail. But then U.S. vice-president Al Gore arrived. Environmentalists adored him because, as shown in his book *Earth in the Balance*, he understood the issues. In 1988, while preparing for the radio series *It's a Matter of Survival*, I had interviewed Gore. I had never heard a politician state the environmental situation so clearly, and he articulated the solutions that were needed.

Much to the disgust of the private U.S. lobbyists, Gore settled for the target of 6 percent reduction in greenhouse gas emissions by 2010. This was in 1997. Even if he had succeeded Clinton as president for two terms, he wouldn't have been in office when the United States was held to account for its compliance, so it could be suggested he had nothing to lose by advancing the deal.

After Prime Minister Jean Chrétien ratified the Kyoto Protocol in December 2002, I was thrilled to receive a letter in January 2003 thanking the foundation for making it possible for him to do so. His letter concluded: "Your personal efforts and those of your foundation have been an important part of the consultation process and have also contributed to informing Canadians about the issues."

Canada's signing was a significant step but did not deliver the numbers needed to make the protocol internationally binding. Russia ratified Kyoto on November 18, 2004, thereby making the protocol international law ninety days later, on February 16, 2005—seven years after it was completed and a mere five to seven years before its end. Unfortunately, under Stephen Harper's Conservative government, Canada withdrew from the Kyoto Protocol in 2011.

Since Kyoto, we've had regular Conference of the Parties summits, including in Paris in 2015, where the Paris Agreement was signed. Kyoto was COP3, Paris COP21, and we're now up to COP31. And yet the most powerful nation on Earth, the U.S., has withdrawn from the Paris Agreement, overturned climate-friendly policies, and is ramping up fossil fuel production. Most countries aren't anywhere near meeting their emissions targets, and the world continues to heat.

We have so many solutions, which carry so many benefits, but we're chained to a system that rewards profit at all costs and so we go on burning fossil fuels despite the certain knowledge that we're heading for catastrophe. It's one of the great disappointments of my life that we in the environmental community and the scientists on whom we depend for the best information haven't been able to get enough people to wake up to the crisis we now face.

Justin Trudeau

PRIME MINISTER
OF CANADA, 2015–25

I have known David for decades now, and for me he has been a teacher, an inspiration, and a friend but also a critic, an antagonist, and a royal pain in the ass.

And I have relished every single moment of it, as his passion for Canada, his love of our planet, and his rigorous scientific mind have always shone through as a challenge to us all to do better and to be better.

Happy birthday, David, and thank you for your extraordinary leadership and service.

With former Prime Minister Justin Trudeau. Theresa Laturnus

Melina Laboucan-Massimo

LUBICON CREE CLIMATE AND INDIGENOUS ACTIVIST, FILMMAKER, PRODUCER

The first time I connected with David Suzuki was over fifteen years ago when he signed an Indigenous solidarity statement for my Nation as we were fighting one of the many pipelines that criss-cross our territory. I knew then that I had found an ally and someone I could trust to stand with not only me and my Nation but with so many Indigenous communities across Turtle Island.

A few years later, David and I spoke to a stadium full of young people in Ontario, which was televised across the province. Following a David Suzuki speech is not an easy task. But because I respected him so much, and to impress him and not fail miserably in front of thousands of people, I rose to the occasion. He helped me grow and become more comfortable with standing in my truth.

David is a mentor and an Elder I look up to. I deeply value the relationship we have shared over the years. I remember several years ago asking him for advice on how to be a good TV show host, as I was in the beginning stages of hosting my own show, *Power to the People*. One of the many things David shared was a reminder to always be myself even on air. And that my voice and opinion, as an Indigenous woman in this world, matter.

One of the biggest things that I admire about David is that he does not mince words even when those words are unpopular. I admire his unwavering integrity and the ways he is always willing to show up, even after so many years of being an incredible advocate for social, environmental, and climate justice.

David is a beacon of hope and an example to never stop sharing what is on your heart by speaking truth to power. Throughout the years I've known David, this has been a constant. His light burns brighter as he continues in what often feels like a dark world where humanity is struggling to find its way.

This has given me courage to do the same and stand in my truth in the protection of Mother Earth and all living beings. I am honoured to know David and his family and be in community and connection with him, Tara, and his daughters Severn and Sarika—who are amazing, brilliant, and beautiful people. What a testament to the extraordinary person David is and what a wonderful father and grandfather!

This world is so blessed to have David in it. I thank him for reminding us how to fall in love with Mother Earth again.

Melina Laboucan-Massimo was the inaugural David Suzuki Foundation fellow; her research focused on climate change, Indigenous knowledge, and renewable energy. Melina is Lubicon Cree from Northern Alberta, founder and executive director of Sacred Earth Solar, and co-founder of Indigenous Climate Action and a TV host.

Ian Mauro

SCIENTIST, FILMMAKER, EXECUTIVE DIRECTOR, PACIFIC INSTITUTE FOR CLIMATE SOLUTIONS AND ENVIRONMENTAL STUDIES PROFESSOR, UNIVERSITY OF VICTORIA

It's the mid-2000s, genetically modified (GM) crops have been commercially released into the environment, and lawsuits are flying all over the Canadian prairies just like the pollen carrying patented genetic material "owned" by the world's biotech companies. Organic farmers in Saskatchewan were planning to sue Monsanto for GM contamination and associated market loss of their crops and invited David Suzuki—a world-famous geneticist—to give a keynote at a fundraiser for their cause. As a doctoral student in my mid-twenties, I had been co-developing research and films with these farmers, and they called me and asked, "Do you want to open for Suzuki and give context about our legal case?" As a young environmental scientist, it was a dream come true to share the stage with Suzuki in support of the agricultural community. After the event, David gave me his card and said, "If you need anything, call me." I never stopped calling, he never stopped being generous with his time, and we have since become friends and colleagues.

I later began making climate change films, and David and I toured many of them to engage Canadians in dialogue and action. I've witnessed David's ability to inspire and transform people's lives, as well as the joy he feels from volunteering his time to cultivate community as a solution to environmental challenges. For many, he's more than the long-time host of *The Nature of Things*; he's one of the narrators of our lives and a beacon for speaking truth to power and uplifting diverse peoples and knowledges to support the ecosystems that keep us alive. When most people his age would have long since retired, he still works harder than anyone I know. His commitment to

the environment and next generations is as legendary as his "cool nerd" look and persuasive voice.

A few years back, David and his wife, Tara, were trying to get from Vancouver to Toronto—without flying—to perform in a play they co-wrote and co-star in called *What You Won't Do for Love*. I volunteered to drive them in an electric vehicle and said, "Let's make a movie." A month later, we were on the road, stopping and filming at various First Nations- and community-led climate initiatives and engaging in hours of car conversation. One moment that stands out: Visiting the Terry Fox Memorial outside Thunder Bay. Seeing the statue, David broke down crying, reflecting on Fox's determination, courage, and ability to inspire hope and national and global unity. It was a window into David Suzuki, a deeply sensitive, passionate, and caring person whose values, tenacity, and belief in the possibility of a safer and healthier world persist despite the adversity we face.

On these journeys with David, it has been clear that as both a geneticist and elder, he has a profound understanding of life and the many lessons it reveals. In a symbolic sense, David's DNA might be considered a double helix threading love and justice, which drives his expression and connection with the ecology of the planet and its people.

Ian Mauro (centre) and Marcel Kreutzer filmed the cross-Canada trip Tara and I took in an electric vehicle in 2022. Janice Williams

STEROCA
3-ple

14 | REFLECTIONS ON SCIENCE AND TECHNOLOGY

TODAY THE MOST POWERFUL FORCE affecting our lives is not politics, business, celebrity, or sports, despite the coverage they receive in the media. By far the greatest factor shaping the world and our lives is science as applied through technology by industry, medicine, and the military. We can't go anywhere on the planet without using the products or encountering the debris of science and technology.

Each innovation changes the way we do things and renders the old ways obsolete. We're seeing ever-more fantastic technologies emerge, such as artificial intelligence and gene-specific engineering, with repercussions we cannot predict. There will also be enormous problems in addition to the ones that already beset us, like global warming, toxic pollution, species extinction, overpopulation, alienation, and drug abuse. Without a basic knowledge of scientific terms and concepts and an understanding of how science differs from other ways of knowing, I don't believe we can find real solutions to such issues.

In my childhood, I wasn't permitted to go to movies or public swimming pools in the summer because my parents worried that I might catch

Starting out as a fruit fly geneticist at UBC, when doing good basic science was all that was required to receive a grant. (Don't be fooled by the lab coat, which I seldom wore.) CBC

polio, a viral disease the Sabin and Salk vaccines later pushed into obscurity. Each year around the world, millions of people suffered agonizing deaths or horrible scars from the now-eradicated disease of smallpox. The world I grew up in lacked jet planes, oral contraceptives, heart transplants, transoceanic phone calls, CDs, VCRs, plastics, photocopying machines, genetic engineering, and so much more. When I began to speak out about the implications of engineering genes, I had no idea of the speed with which techniques to manipulate DNA would develop.

Not only does each innovation alter the way we do things, but some also change the very definition of what it is to be human. We love technology because we design it to do specific things for us; we seldom reflect on the consequences or have any inkling of what the long-term repercussions might be. Thus, we discovered biomagnification of pesticides, the effects of chlorofluorocarbons (CFCs) on the ozone layer, and radioactive fallout from nuclear weapons only after the technologies had been created and used.

When I began my career as a scientist, we took pride in exploring basic ideas of the structure of matter, the origin of the cosmos, or the structure

Jim Fulton and I present Prime Minister Paul Martin (centre) with our report *Sustainability Within a Generation* in 2004.

and function of genes without having to justify the expansion of human knowledge. Medical genetics was considered intellectually inferior to the kind of work we carried out with fruit flies.

We support science because it is a part of what it means to be civilized, pushing back the curtains of ignorance by revealing bits and pieces of nature's secrets. But more and more, we are under a demand that science deliver practical uses. This is a dangerous requirement, because it imposes an urgency that can lead to shortcuts, unwarranted claims, and deception.

Since I wrote *Metamorphosis: Stages in a Life*, published in 1987, I have abandoned practising genetics, which had consumed me for more than a quarter of a century. In the 1970s, when geneticists began to learn to isolate and manipulate DNA in very sophisticated ways, it was immediately obvious there were enormous social, economic, and ecological implications. For decades, writers, philosophers, and geneticists had been speculating about genetic engineering and discussing the potential ramifications of such powers. Having belatedly recognized the dangers that our inventiveness posed from the battles over the insecticide DDT and then CFCs, I felt genetic engineering would encounter the same problems—our manipulative powers were great, but our knowledge of how the world works is so limited that we would not be able to anticipate all of the consequences in the real world.

As a result of the grotesque misapplication of a genetic rationale during the early part of the twentieth century in the eugenics movement, and in the Japanese Canadian evacuation, and then in the Holocaust, I knew a debate about genetic engineering had to be engaged, and I wanted to participate in it with credibility. So I began to write a series of disclaimers, stating my intent not to become involved in such research, even though it was perhaps one of the most exciting moments in the history of genetics. I wrote a series of columns that led to my eventual withdrawal from research to maintain my credibility in the discussions about the implications.

Nevertheless, I continue to take vicarious delight in the enormous technical dexterity of today's molecular geneticists and revel in seeing answers to biological questions I never thought would be resolved in my lifetime. I watched my daughter Severn carrying out experiments in undergraduate labs at Yale University that were unthinkable when I graduated with a PhD. Still, the rush to exploit this new area as biotechnology has me deeply disturbed.

As a geneticist, I believe there will be monumental discoveries and applications to come. I also know that it is far too early and that the driving force behind the explosion in biotechnology is money. Though I deliberately stopped research, I did not immediately lose all the knowledge that made me a geneticist. I am proud of my career and contribution in the field, yet the minute I ceased doing research and began to speak out about the unseemly haste with which scientists were rushing to exploit their work, people in biotechnology lashed out as if I lacked credibility because I was no longer up-to-date enough to understand what was being done.

With Naomi Klein (left) and Melina Laboucan-Massimo at a 2018 David Suzuki Foundation fellowship reception in Toronto. Robert Elliott

The power of science is in description, teasing out bits of nature's secrets. Each insight or discovery reveals further layers of complexity and interconnections. Our models are of necessity absurdly simple, often grotesque caricatures of the real world. But they are our best tool when we try to "manage" our surroundings. In most areas—such as fisheries, forestry, and climate—our goal should be to guide human activity. Instead of trying to bludgeon nature into submission by the brute force applications of our insights (if planted, seedlings will grow into a forest; insecticides kill insects), we would do better to acknowledge the 3.8 billion years over which life has evolved its secrets. Rather than overwhelming nature, we could try to emulate what we see. Biomimicry should be our guiding principle.

But even reductionism—focusing on parts of nature—can provide stunning insights into the elegance and interconnectedness of nature and reveal the flaws in the ways we try to manage it.

Speaking at the 2019 youth climate march in Vancouver. Janice Williams

Nancy J. Turner

ADJUNCT PROFESSOR EMERITA,
SCHOOL OF ENVIRONMENTAL
STUDIES, UNIVERSITY OF VICTORIA

David Suzuki is a hero of the planet. Anyone anywhere who is concerned about the future of Earth and all its life forms, or who just has an interest in the natural world, knows about, admires, and follows his work and his words. I have had the great pleasure of witnessing the adoration he receives, even to his embarrassment at times. Around 2020, David and Haida leader Miles Richardson, who have been friends for a long time, conceived of a project they called Reconciling Ways of Knowing, which would bring together Western scientists and other academics with Indigenous Elders and knowledge-keepers, holders of traditional ecological knowledge and wisdom. David, as a geneticist with a strong background in science, was drawn to the place-based, grounded knowledge of Indigenous Peoples. He had spent time in many First Nations communities, especially in British Columbia, and had many friends in places like Bella Bella, Hartley Bay, and Haida Gwaii. Miles, a long-time member of the David Suzuki Foundation board, recognized the significance of the scientific method along with his own Haida teachings.

David and Miles invited friends and colleagues with similar interests to participate in their endeavour, and I was fortunate to be one of them. That's how I found myself sitting in a restaurant in Winnipeg, en route to Turtle Lodge, where the inaugural meeting of the Reconciling Ways of Knowing group was to be held, hosted by Nii Gaani Aki Inini Dave Courchene, respected and loved Elder and knowledge-keeper of the Anishinaabe Nation. Sitting with David at the restaurant, I was witness to the looks of astonishment and adoration from the people around us, as it dawned on them that their hero was there in the same room. Most were shy about approaching, but two teenaged boys came up to David and asked, "Are you *really* David Suzuki?" They were ecstatic! He signed his autograph for them, and they left, happy to have had such a once-in-a-lifetime experience. I can imagine them talking about meeting David Suzuki to their own children and eventually to their grandchildren.

It's not just the adoration that David engenders; it's a realization that he brings to us of the importance of the environment and of the need for many different approaches to understanding our planet. At the Turtle Lodge gathering, the scientists, including David and me and the late brilliant limnologist David Schindler, were adopted formally by Elder Courchene and the other Indigenous knowledge-keepers. That was a special and touching moment. After that meeting, because of COVID-19, the Reconciling Ways of Knowing forum was converted to multiple online sessions extending over the next two years. (Anyone with access to the web can watch these dialogues at rwok.ca.) We are working to have these published in book form, which will be another lasting tribute to a truly great man!

John Lucchesi

GENETICIST, PROFESSOR EMERITUS OF BIOLOGY AT EMORY UNIVERSITY

I met David in 1965. I was a postdoctoral fellow at the University of Oregon in Eugene, and my mentor took us to Seattle where a famous geneticist, James Crow, was giving a seminar. David and his students had come to Seattle for the same occasion. Four years later, when I was on the faculty of the University of North Carolina, Chapel Hill, he invited me to teach a course and spend the summer in his new lab at UBC. My wife, Bobbi, and five-year-old Annalena drove to Vancouver, but we flew our German shepherd. David had a field day. He told me that it was unconscionable to spend money on trivial matters when there was so much misery and hunger in the world. My dog could not understand him, but I walked around with my tail between my legs... until he bought a new TV set to watch the moon landing in colour! I told him that it was unconscionable to spend money on such a trivial matter since he had a perfectly good black-and-white set. He smiled broadly, suggesting to me that we were even.

The three months that I spent in his lab constituted one of the most significant experiences that influenced me personally and scientifically for the rest of my life. Our unifying bond was scientific research. We all talked about it almost constantly. We went fishing together, camping together, had dinners together, and smoked a little pot together (but never inhaled, of course).

As a super young scientist, David soon began the personal transformation that led to the vocation that has dominated his life. First, he began to investigate if scientists were conscious of and felt responsible for the social consequences of their discoveries (among the people he interviewed was Edward Teller, considered the father

of the hydrogen bomb). David's social consciousness led him to a vivid concern about the effect of human activities on populations and their environment. (He admires and loves Indigenous Peoples because of their exemplary relationship with nature.) I followed his personal evolution and the amazing steps he has taken to make a difference. Although, as with all visionaries, he has had moments of disillusionment on the progress he has made. I said to him that his success should be measured by the recognition and affection young Canadians have for him and the fact that his name is known in all Canadian households. Over the past sixty years, I have been grateful for David's close friendship and for the enrichment that it has brought to my life.

With friend and fellow geneticist John Lucchesi on Haida Gwaii.

Bill McKibben

AUTHOR, EDUCATOR, ENVIRONMENTALIST, CO-FOUNDER OF 350.ORG

David is a pioneer in so many ways—as a communicator above all, perhaps. But for me, he represents something else as well: a very, very early example of a gifted bench scientist being willing to engage in full-throated advocacy for the things he knew were important. It's a trend that's grown in recent years, but we should never forget that David was willing to take this step when it was controversial and hard. A Paul Revere in so many ways!

Campfire at Beatrice Lake in the Slocan Valley. This is my happy place, in spite of our incarceration. From *Force of Nature: The David Suzuki Movie*. Ari Gunnarsson

15 | A CULTURE OF CELEBRITY

IT IS ASTONISHING AND FRIGHTENING to see the extent to which the phenomenon of celebrity has come to dominate our consciousness. When the media lavish as much attention (or even more) on celebrities as they do on weightier issues, how can people distinguish what is important from what is not? The result of our preoccupation with celebrity is that the opinion of someone who might be a lightweight or a fool carries as much heft as the words of a scientist, doctor, or other expert.

We live in a time when the military, industry, and medicine are all applying scientific insights, with profound social, economic, and political consequences. As a result, ignoring scientific matters is dangerous. It's not that I believe science will ultimately provide solutions to major problems we face; I think solutions to environmental issues are much more likely to result from political, social, and economic decisions than from scientific ones. But scientists can deliver the best descriptions of the state of climate, species, pollution, deforestation, and so on, and these should inform our political and economic actions.

I never sought or desired celebrity, but television provides a kind of intimacy that movies do not. So when people run into me, they often greet me as a familiar friend. I can't help being startled each time someone

With famed Oneida actor Graham Greene (1952–2025) in a gag shot.

addresses me, though almost always it is to say something kind. I must admit I am not able to respond generously because greetings are still a surprise and intrusion, and my teenage reticence to engage in conversation returns.

When the CBC touted its search for "the greatest Canadian" for a television series of that name in 2004, I was interviewed on radio and asked what I thought about the idea. I scoffed at the notion that it meant anything. Besides, how can we select one person out of millions who are Canadians and conclude that one individual is the greatest?

I now realize that the exercise of trying to define the greatest Canadian was not a wasted or even frivolous effort. I was astonished to watch and listen to conversations, often quite heated, about Canada and Canadians. It was great to hear the talk and feel the passion—it got us thinking about this country, its values, and what makes us special.

As a scientist and an environmental activist, I had no idea that I would be anywhere on the list, so when the names were first announced, I was surprised to be placed among the top ten. I would have been honoured to be in the top one thousand. What a remarkable list—starting at ten with hockey great Wayne Gretzky, then scientist and inventor Alexander Graham Bell, Canada's first prime minister Sir John A. Macdonald, hockey commentator Don Cherry, Nobel Peace Prize laureate and Canada's fourteenth prime minister Lester B. Pearson, me, insulin discoverer and Nobel laureate Sir Frederick Banting, Canada's fifteenth prime minister Pierre Trudeau, athlete and activist Terry Fox, and former Saskatchewan premier and New Democratic Party leader Tommy Douglas at number one. Douglas was a socialist who championed national medicare and many other social causes. Would such a person have even appeared among the

The Mounties always get their man. At the Banff World Television Awards.

TOP: With Jane Fonda and Tom Lovejoy at a conference in Malibu, California. Lance Webster BOTTOM: With Ed Begley Jr. and Jane Fonda at my eightieth birthday. Brendon Purdy

top one hundred Americans? I felt our list alone indicated how Canada is different from the U.S. But it also showed a blindness: no women!

Only a few weeks later, *Maclean's* magazine published the results of a poll in which women across Canada were asked with whom they would most like to be stranded on a desert island. They were asked to select from a small list that included me, CBC Television newsreader Peter Mansbridge, Canadian prime minister Paul Martin, *Canadian Idol* host Ben Mulroney, and Calgary Flames ice-hockey superstar Jarome Iginla. I was flabbergasted when a writer with the magazine called to tell me I had been selected first, by 46 percent of the women (55 percent in Alberta!), while the runner-up was young Mulroney at 16 percent!

LEFT: **William Shatner interviewed me in my home in 2017 for the documentary film *The Truth Is in the Stars*.** RIGHT: **With Sarah Harmer, a musician and fighter for the environment and social justice, at the If You Were Prime Minister Tour, 2007.** Theresa Laturnus

Neil Young

SINGER, SONGWRITER, ACTIVIST

Thanks for being there for Canada all these years. You have always been so supportive of Canada, and we all appreciate that greatly.

As the years pass, Canada becomes more of a jewel of a nation. I am so proud to be Canadian and be your friend.

Neil Young and his band were a highlight of the David Suzuki Foundation's 2014 Blue Dot Tour. Kris Krüg

Melissa Lem

PRESIDENT OF THE CANADIAN ASSOCIATION OF PHYSICIANS FOR THE ENVIRONMENT

The first time I met David was not my finest hour. Early on in my environmental advocacy journey, I penned a "Docs Talk" blog post for the David Suzuki Foundation about nature's health benefits. When an invitation later came to his VIP reception at the Art Gallery of Ontario with nature-reconnection author Richard Louv, I was outrageously excited. Growing up in a majority-white suburb of Toronto, I had found refuge from racist bullying in my neighbourhood green spaces—and by watching David, an Asian Canadian like me, light up my television with the same care for the planet and love for storytelling I held so deeply. In my pantheon of "If you could meet one person in the world, who would it be?" luminaries, he was twinkling brightly at the top.

I gathered up my courage at the AGO, cornered him, and nervously blurted out how I was a huge fan and had followed his career my entire life. He smiled and nodded as I receded back into the hubbub of admirers, bereft of further words. I'd fumbled the pass—the one chance I might have had to have a real conversation with David Suzuki.

Today, almost fifteen years later, I wish I could tell that blushing, crestfallen doctor we would go on to have many more conversations. As president of the Canadian Association of Physicians for the Environment, I've been buoyed by his steadfast support of our mission to protect people's health by protecting the planet. Like us, he's convinced that doctors are some of the world's most powerful messengers for planetary health. Whether it's lending his celebrity to our events—invariably bringing down the house—or plotting campaign strategy in his living room while my son buries his nose in a book on the couch, I leave every encounter with some valuable new insight or action item, the path forward clearer.

But it's the times when our clans come together in a maelstrom of nature-infused fun that enliven these critical years. David and Tara's cabin has become a waypoint for memories during my family's island travels for my rural medicine work: Sarika's kids lobbing pine cones at mine in the mossy woods, digging spouting clams and oysters out of wet sand in the sunshine, and a surprise homemade chocolate cake blazing with candles on my birthday as rain drums down on the roof.

The memory I cherish most, though, is when David taught my son how to fish—something I grew up doing with my own family in the waterways of Ontario. In full grandpa mode, everyone clicked into bright red life jackets, David navigated his boat into just the right blue depths of the Salish Sea and patiently taught my son how to jig a lure as others wielded their fishing rods like it was second nature. In that full-circle moment, I thought of the adage "Give a man a fish and you feed him for a day; teach a man to fish and you feed him for a lifetime," as I watched one of my childhood heroes, now my friend, making sure my son felt a sense of belonging outdoors as he learned a new skill.

Thank you, David, for your shining legacy of land, water, air, and biodiversity protection, telling the stories about the environment we all need to hear, and empowering us to do good by example. Thank you for being an unbreakable line relentlessly pulling this ecosystem of tree-hugging outsiders, warriors, and dreamers for a healthier planet of every generation together out of the deep and into the light. Thank you for teaching us all how to fish.

16 | THOUGHTS AS I GROW OLD

IN 1986, I RECEIVED the Royal Bank Award for Canadian Achievement, a tax-free $100,000, with which I bought our beloved Tangwyn, a small piece of paradise on Quadra Island. When we finally purchased *Kingfisher*, a small cabin cruiser, in 2003, Tara proclaimed, "David, we've got everything we need in life. We don't need any more stuff."

As we ride the last ferry from Campbell River to Quathiaski Cove on Quadra, our excitement grows and we delight in the sight of the island hills covered in forest, the dense schools of herring, the fishing boats pursuing salmon, and the ever-present eagles ready to swoop down and take a careless fish.

But when we talk to our neighbours who have lived in the area for fifty years or more, they describe a world that no longer exists around there: bays filled with abalone, red snapper, gigantic ling cod and rock cod as long as an arm, herring so abundant they could be raked off the kelp to fill a punt in minutes, and schools of salmon so thick they could be heard coming as their bodies slapped the water kilometres away. Today all of that is gone.

Enjoying peace at our cabin.

When I was growing up in Vancouver, Dad would row a boat around Stanley Park in downtown Vancouver and catch sea-run cutthroat trout. We would jig for halibut off Spanish Banks on the city's waterfront, catch sturgeon in the Fraser River, and ride horses up the Vedder River to catch steelhead and Dolly Varden trout. My grandchildren have no hope of experiencing the richness I knew as a child.

Being a parent is the most important and joyful thing I have done in life, and I have always been committed to my children, though not in the same way my father was. After my first marriage ended, I endeavoured to be with the children every day I was in Vancouver and was aided by Joane's generosity in allowing me access to them without complaint or restriction. But often my mind was distracted, not totally focused on them but off somewhere else. I was too selfish to give myself over to being Dad 100 percent, and I regret that.

Joane was my first love, and she has always had my greatest respect and gratitude for tolerating my shortcomings and for never using the children as a weapon to punish or hurt me. Faculty members could

The family at Tangwyn.

enroll their children in university tuition-free, but I told my children I would only cover their postsecondary education if they got it outside British Columbia. Being free from home, meeting new people, establishing relationships, and making mistakes as adults are half the university experience.

Tamiko studied biology at McGill University in Montreal. There, she fell in love with Eduardo Campos, a Chilean Canadian who was enrolled in engineering and was a computer whiz. They married after graduation and decided to have a footloose life, working for periods and saving enough to travel to different parts of the world. In 1990, Tamiko gave birth to Tamo, my first grandson, and three years later to Midori, my first granddaughter. Eduardo took jobs working in South America and spent a lot of time away from home. Sadly, in 2011, he was killed by a hit-and-run driver in Guatemala while travelling by motorcycle on his dream trip from Santiago, Chile, to Vancouver. Months later, Tamiko held a meeting of his friends and family in Lighthouse Park near Vancouver. She told us to take a handful of Eduardo's ashes and put them somewhere he would have

Laura, Tamiko, and Troy.

liked. I scrambled down rocks to a pool by the beach and threw the ashes onto the water. Immediately small fish rose up to feed on the ashes. He was back in nature's cycle of life.

Tamo and Midori were born when Sarika was still a child, so suddenly I had a young daughter and grandchildren when I was spending a lot of time away. I loved attending basketball games to cheer Sev and Sarika when they played in high school but have seldom been in town when Tamo and Midori have had hockey, soccer, snowboarding, and football competitions.

Tamo is now an environmental activist in his own right and abandoned his hopes for a snowboarding career after enduring too many injuries. He has made several powerful films on environmental and Indigenous issues. His sister, Midori, lives in Masset on Haida Gwaii, where she works with Haida youth. She and her Haida partner will pass on some of my genes with a great-grandchild and tighten my ties to Haida Gwaii.

I had thought having children was the most wonderful gift in my life until they gifted me with the greatest joy, grandchildren. They are such a

LEFT: **With grandson Tamo Campos, an environmental activist, filmmaker, and former professional snowboarder.** RIGHT: **Son-in-law Peter Cook with Laura and my grandson Jonathan.**

delight because the relationship is so different from the relationship with one's children. Every human relationship—between lovers, parents, or children—has moments of frustration, anger, and resentment. With grandchildren, however, there isn't the chafing that can result from living together day in and day out, so every get-together is a celebration and joy.

Laura chose to attend Queen's University in Kingston, Ontario, where she majored in psychology. I was delighted when she fell in love with and later married Peter Cook, a fellow cartoonist on the school paper and a psychology major. Jonathan, their son, is a beautiful child who was found to have suffered oxygen deprivation at birth and has cerebral palsy. What has been so impressive and humbling to me has been the parenting of this heroic young couple.

Troy spent many years trying to figure out his relationship with me, but he stayed very close to my father, moving in with him for several years. We have become close again. (Thank goodness for email.) Like many younger men today, he has chosen not to follow the high-pressure, competitive path that was the model of a "successful" man when I was younger. As a result, in so many ways, he has led a more varied, interesting life than I have. After graduating from Emily Carr University, he ended up in Dawson City, Yukon, where he has lived for years as a carpenter and filmmaker. He and his partner, Molly MacDonald, live comfortably in a house he built and have two beautiful children, Setsu and Kai, who are genuine Yukoners.

After graduating from Yale University in 2001, Severn travelled for two years and gave inspirational speeches to adult and youth groups across

With Troy in the hull of the *Klondike*, a sternwheeler boat in Whitehorse, Yukon, that Troy was helping to restore.

North America. She then decided to go back to graduate school to study ethnobotany. She became executive director of the David Suzuki Foundation for several years and later returned to the University of British Columbia where she completed her PhD in Haida language revitalization.

Sarika decided to go the University of California at Berkeley where she earned a degree in marine biology. She went on to earn an MSc in marine biology with Daniel Pauly at UBC and a PhD from the University of York (U.K.) with Callum Roberts. She has since taken over as co-host, with Anthony Morgan, of *The Nature of Things* after my retirement in 2023 from forty-four years of hosting the show!

All my children have become vibrant, interesting human beings, all of them committed environmentalists and contributors to society. If my children and their children know anything, I hope it is that they have my unconditional love and can always depend on that.

Christmas 2003. Back row, L-R: Tamiko, Midori (my granddaughter), Sarika. Front row, L-R: Severn, Huckleberry, Tara, Eduardo Campos (Tamiko's husband), Tamo (my grandson).

Human beings bear that terrible burden that self-awareness has inflicted on us—the knowledge that we, like all other creatures on Earth, will die. Belief in a life after death is one way to bear this truth, although it pains me to see people who seem to care little about this life because they believe they will live forever after they leave it.

As an atheist, I have no illusions about my life and death. As I ponder infinite time and distance in the universe, I and all of humanity are but a momentary flash, insignificant in cosmic terms. Human time and distance are irrelevant in the context of eternity. We, like all species, will become extinct. In the end, as we reflect on the meaning of our lives and our legacy to the future, what more could we ask for than to be remembered with affection and respect by a few people who will survive a decade or two further, by our children and grandchildren? I hope when it's my time to die, I do so with the dignity of my father.

I moved in with Dad to be with him in his final weeks of life. He was still alert and interested in what was happening in the family. Each night, Tara and the girls would come over, and they sometimes brought slides of one of our trips, often ones we had taken with Dad and Mom.

My favourite photo of Dad in repose at Windy Bay, Haida Gwaii. Jeff Gibbs

In the last week, my sisters arrived, and we reminisced about our lives. What struck me was that at no point did we complain about how hard life had been or all the things we had missed out on. Instead, we laughed and cried over stories about family, friends, and neighbours and the things we had done together that had enriched our lives. There was no boasting about possessions or wealth or accomplishments, only human relationships and shared experiences, which are what life is all about.

Perhaps one or two programs I've done on television or radio will be played again after I'm dead, perhaps a book or two I've written will be read. That would be nice. But the one true legacy of any value is my children

Sharing my love of washing dishes with my grandson Carr.

and, through them, my grandchildren. I now have ten grandchildren, and each one adds to my joy. My grandchildren may remember something they learned from me or shared with me, and if I'm lucky, they may even pass that snippet on to their grandchildren. So at most, I might be remembered for four generations.

This is a sad time to depart from this life. I have witnessed the disappearance and destruction of so much of the natural world that I loved. Our trajectory to dominance of the planet has been spectacular, but we have not fully comprehended the price of that success. It has been my lot to be a Cassandra or Chicken Little, warning about imminent disaster, and it gives me no satisfaction at all to think my concerns may be validated by my grandchildren's generation.

My grandchildren are my stake in the near future, and it is my most fervent hope that they might say one day, "Grandpa was part of a great movement that helped turn things around for us." I also hope that they might remember my most valuable lesson and be able to say, "Grandpa taught us how to catch and clean a fish. Let's go catch one for dinner."

Anthony Morgan

CO-HOST, CBC'S
THE NATURE OF THINGS

The first time I met David Suzuki was at his home in Vancouver. His daughter Sarika and I had just officially taken over hosting *The Nature of Things*, and we were filming a kind of symbolic hand-off—the passing of a torch from him to us.

His home was exactly what you'd imagine from a man who spent his life travelling the world, asking questions, and meeting people. There were masks, carvings, dried plants, and photographs everywhere—a bit dusty but clearly cherished. You could see that everywhere he went people wanted to give him something to say thank you. He was calm in his home, in his element. It was a rainy Vancouver morning, and he was dressed like David Suzuki always is: hiking hat, outdoor gear. I can't say I'm positive he was wearing denim, but the odds are good. He was warm, welcoming, and full of stories. I remember thinking, "This is what it feels like to visit a cultural icon."

Inside, I felt a mix of emotions. Honoured, nervous, giddy even. And deeply aware of the responsibility I was being asked to shoulder. You don't want to make a fool of yourself in front of David Suzuki. You want him to believe that maybe, just maybe, you might be worthy of the job. But how do you live up to David Suzuki? The man fundamentally changed the way Canada thinks about the environment—and our role in protecting it.

And yet I could feel he was rooting for me. He wasn't trying to intimidate or control the future of the show; he wanted to guide, to support. In that short afternoon, he gave me three pieces of advice I carry with me to this day. First, that the stories are what matter, not

the host. Second, to stay close to my curiosity—it's what got me here. And third, to embrace collaboration while staying grounded in integrity. That last one defines him best. David Suzuki has always spoken the truth, even when it wasn't popular. He never shied away from difficult conversations—even when asked to play it safe in shifting cultural moments. He wasn't interested in optics. He was interested in doing what was right.

And he didn't take himself too seriously either. I remember Sarika and I chatting with him in his living room. Looking through photos, I was astonished to realize that an infamous photo of David posing nude with just a leaf covering his sensitive bits... was real! I had assumed it must have been AI-generated—the internet being the internet. But no. The man was shredded. I remember the three of us dying laughing as Sarika described her school friends talking about how jacked her dad was.

That's the beauty of David Suzuki—a life of courage, guided by curiosity, grounded in humility, and open to joy. Big shoes to fill, but great steps to follow in.

I was happy to see *The Nature of Things* in good hands with my daughter Sarika and Anthony Morgan taking over as hosts. L-R: me, Sarika, Anthony. CBC Licensing/Jimmy Jeong

Sarika Cullis-Suzuki

CO-HOST, CBC'S
THE NATURE OF THINGS

It's interesting to think of the evolution of a relationship. For me and my dad, it was as child/parent, teen/parent, parent/grandparent. And through each very different phase, Dad's always been there for me.

Dad showed me how to catch and clean my first fish, how to give a speech, and how to shoot a basketball. He passed on his love of chocolate cake, camping, and berry picking. He taught me how to dissect roadkill, pin dead insects, travel, and fly-fish. He taught me how to listen to others and that every person has an interesting story to tell. He taught me the love of "goss." He taught me the magic of swamps, the importance of working out, and that there are no short-cuts when it comes to working hard. He also taught me, when I was stunned speechless after learning I was pregnant with twins, that sometimes we can't control the best things in life.

For better or worse, I've inherited Dad's "no filter" gene when it comes to saying what's on my mind. He feels things with an intensity and passion that has never faded . . . and yet at night, right before bed, when the work and computer go away, and the news is turned off, he becomes a hilarious, carefree person and is always telling me to relax! We've always been close, but as Dad completes his ninetieth year, I am filled with a gratitude that overwhelms, in particular for the past ten years, the time we've spent together since I had kids. I've been able to get to know him as a grandfather to my children. When I'm away for work, he will often come over to help out, picking them up after school, getting their homework assignments, reading them

bedtime stories, making them breakfast, taking them to soccer practice or to a nearby swamp or to the insect zoo. But mostly he is *there* for them.

They see you being you, Dad, building treehouses, writing articles, doing the dishes, speaking up for what you believe in, talking about injustices—and they will remember that for the rest of their lives. You are etched into their beings forever, dear Bompa. They adore you. We all do. We are so unbelievably lucky. I love you, Dad.

Family gathering in June 2005. Back row, L-R: Severn, Tara, Peter Cook (my son-in-law), Laura, Delroy Barrett, Jill Aoki (my niece), Makoto (Marcia's grandson). Front row, L-R: Sarika, me, Jonathan (my grandson), Marcia, Richard Aoki (Marcia's husband), Malevai (Richard's grandson).

John Ruffolo

FOUNDER AND MANAGING PARTNER, MAVERIX PRIVATE EQUITY

My relationship with David Suzuki is deeply meaningful and influential. To me, he represents a guiding voice—one that has challenged me to think critically about the world I inhabit and my responsibility within it. His lifelong dedication to environmental advocacy, science communication, and Indigenous rights has shaped the way I understand the intersection of nature, society, and justice.

As I was growing up, David Suzuki was a familiar name in my household. His documentaries and public broadcasts weren't just informative, they were transformative. They introduced me to the fragility of our ecosystems and the urgency of the climate crisis long before these topics were widely accepted in mainstream discourse. I remember watching *The Nature of Things* and feeling a mix of wonder and concern: wonder at the beauty and complexity of the natural world, and concern over how easily that balance can be disrupted.

What stands out most to me is David's ability to blend science with humanity. He doesn't just present facts; he connects them to our values, cultures, and responsibilities as stewards of Earth. His work is about protecting the environment, about protecting the future of life itself. That framing has pushed me to reflect on my own choices—what I consume, how I live, what I support—and to consider the broader impact of those decisions.

Beyond his environmentalism, David's courage in confronting political and corporate interests, even when it's unpopular or inconvenient, is admirable. His authenticity and unwavering voice remind me that advocacy isn't about being perfect—it's about showing up, staying informed, and being consistent in your values. In a world increasingly shaped by short-term gain, David's long-term vision is not just refreshing, it's vital. He's given me the room to also say things that might be deeply unpopular.

My relationship with David is, in many ways, both aspirational and personal. He embodies the kind of integrity and purpose I strive to bring into my own life. He's a reminder that science, when grounded in compassion and courage, can be a force for transformative change. He is also family. I have travelled with him in our great country and in far-flung places around the world. He is a mentor to not only me but also to my spouse, Carryn, and children Caymus and Rome.

David Suzuki represents the voice of conscience in a noisy world. To me, he's more than an environmentalist—he's a mentor, a symbol of hope, and a call to action. David is family.

Canadian hero Rick Hansen (centre) loves fishing and took John Ruffolo and me out on his boat.

17 | REFLECTIONS ON LESSONS LEARNED

HAVING CIRCLED THE SUN NINETY TIMES, I have achieved the status of elder. I believe my legacy as such is not material, but the lessons I've learned over a lifetime of experiences and thought are. Family has been and continues to be the greatest joy, stress, and motivation in my life—from my parents and siblings to my two wives and our children and their children. Community, students and researchers in my university lab, and staff, funders, and supporters of the David Suzuki Foundation have also helped guide my path.

Two major issues have consumed most of my adult life. The first is social justice beginning with my experiences as a Canadian of Japanese origin during the Second World War and then in Oak Ridge, Tennessee, where I witnessed discrimination against my lab partner, who was Black. The second issue is the dominant ideology that perpetuates a disconnect between humans and the natural world, promoting a belief in our supremacy without restraints of natural laws, with catastrophic consequences.

For almost all of human existence, our species understood that we are one strand in a complex web of relationships with all other life and with air, water, soil, food, and sunlight. Indigenous land defenders understand

My grandkids have helped me to remember what life is like through children's eyes.

this "ecocentric" perspective. In a web, every action has consequences and therefore we have responsibilities to "act in a good way" to ensure nature's continuing abundance and generosity. That has been our understanding for almost all of human existence.

Suddenly, in evolutionary terms, humanity embraced an "anthropocentric" perspective, which posits humanity astride a planet that is ours for the taking. We believe we can live this way because we're smart and needn't worry about "planetary boundaries"—such as atmospheric carbon levels, ocean pH, and so on—that constrain all other species.

What we have learned in a mere few centuries is remarkable: what stars are, what causes earthquakes and volcanoes, that there is energy within atoms, how sex is determined, and what DNA does. These are incredible discoveries, but acting on our knowledge has had enormous consequences, positive and negative.

Now, in the face of overwhelming evidence of a catastrophic loss of species, human-induced climate change, and plasticization of the planet, we continue to support the legal, economic, and political systems created to shape, guide, and constrain humans. They have become agents of environmental degradation because they ignore nature as the foundation of our existence and therefore fail to guide our activities around nature's continued well-being.

Giving a speech at the Chan Centre at UBC for the film *Force of Nature*.

We can and must change our behaviour and the systems that support us. Unless we rein ourselves in and constrict our actions, our attempts to "manage" any part of nature or "mitigate" the enormous consequences are doomed to fail because of the scale and speed of corrective measures needed. We have overshot safe boundaries. Furthermore, impressive as science has been, the planet is far more complicated than we understand. We still have much to learn about its geophysical, chemical, and biological components and how they interconnect and interact.

The only factor we have any possibility of managing is the top planetary omnivore, us.

When European explorers began to search for new resources, they found the world was already fully occupied and developed by a vast array of plants and animals that sustained diverse peoples for tens of millennia.

Gardening is a magical activity filled with the wonder of life. Let us have gratitude for the seed that grows into food.

The new arrivals were driven by a lust for new lands, opportunities, and riches that they "discovered" in Africa, Asia, North and South America, Australia, and New Zealand. But Indigenous Peoples had a different relationship with their homelands and so were seen by Europeans as impediments to the search for wealth and opportunity. Genocidal policies and practices through conquest and colonization enabled exploitation by virtually every European country. Thus, a white European resourcist or extractivist world view spread around the globe.

IN THE PAST CENTURY and a half, humanity has undergone a profound shift from rural villages to cities, where our primary focus is our job. We assume nature is "out there," away from us and protected in parks and

Sea Shepherd Conservation Society news conference, 2016, aboard the *Martin Sheen* to launch Operation Virus Hunter, highlighting fish farm dangers. L-R: Kwikwasut'inuxw Haxwa'mis Chief Bob Chamberlin (chair of First Nation Wild Salmon Alliance), me, Alexandra Morton, Stó:lō Chief Ernie Crey, Sea Shepherd co-founder Rod Marining, Pamela Anderson.
Sea Shepherd Conservation Society

reserves. This is a fatal mistake. Even in the biggest cities, nature surrounds us and is in us.

We cannot survive without air for more than three or four minutes. Air keeps our lungs from collapsing when we exhale, while every heartbeat delivers oxygen to all parts of our bodies. Pure oxygen-rich air is not created by human factories or machines but by all marine and terrestrial plants. We are air yet we dump toxic chemicals from cars, factories, and jets into it as if they will not affect us.

In the same way, we have an absolute need for rich soil to grow food and pure water to inflate our cells. All of the energy we need to move, work, and play is sunlight captured by photosynthesis.

Earth (soil), air (atmosphere), fire (sunlight), and water are captured, recycled, cleansed, and replenished by nature—sacred gifts that are the foundation of our existence and well-being.

In 1960, the eminent population geneticist Richard Lewontin came to the zoology department of the University of Chicago where I was doing my PhD research. With Jack Hubby in our lab, he analyzed proteins specified by a number of genes in individual fruit flies. They made a huge discovery: a high degree of variation of each gene in the population. Now we know this is part of nature's survival strategy because diversity provides more options for selection when environmental conditions change.

In ecosystems, species diversity provides greater resilience in the face of floods, storms, fires, and drought. Diversity among ecosystems enables life to flourish in many different environments, from Arctic tundra to tropical rainforests, deserts, and wetlands. Like Russian dolls, diversity

In 2019, I was awarded a Right Livelihood Award—the "alternative Nobel"—along with Congolese environmental activist René Ngongo (left) and New Zealand peace activist Alyn Ware.

is built into gene, species, and ecosystem levels to maximize resilience and adaptability over time.

Humans create place-based culture, and its wonderful diversity has enabled our species to flourish everywhere. Wade Davis calls the sum of all human cultures the "ethnosphere." In the biosphere, loss of species reduces resilience. So, too, in the ethnosphere: Loss of language and culture catastrophically diminishes humanity's adaptability.

Over and over in human activities, diverse perspectives and ideas provide maximal choices out of which solutions may be found. In the United States, immigrants are accepted into a "melting pot" of American culture and values that celebrate homogeneity. In contrast, former Canadian prime minister Pierre Trudeau introduced the model of a "cultural mosaic" wherein immigrants from diverse cultures are valued for what they bring to Canada. Good for Canada.

If you autopsy a thousand bodies from all over the world, examining only the sense organs, neurons, and brains, you could not identify Black, white, or Asian bodies because we are all built on the same basic plan. We are one species. In any given situation, our sense organs detect the same sounds, smells, tastes, vision, and touch, but the incoming information is edited through filters shaped by gender, ethnicity, religion, language, culture, and socio-economic status. That's why eyewitness accounts of an event are so notoriously unreliable. While we may share witnessing it, what we actually perceive and remember will differ from person to person.

In April 2019, I had a video call with Canadian astronaut David Saint-Jacques while he was on board the International Space Station.

Diversity of information, ideas, and opinions is essential to solve problems confronting us, but how can we find agreement on solutions when we can't even agree on whether there is a problem, how serious it is, or even its cause? Science is one source of information in which all participants recognize and agree on methodology, observation, and results. In 1992, fifteen hundred scientists from seventy-one countries, including more than half of all Nobel Prize winners, signed the famous "World Scientists' Warning to Humanity." In 2017, fifteen thousand scientists from 184 nations signed "Warning to Humanity: A Second Notice." As scientists from a wide spectrum of cultures, they were in complete agreement, and that's what's desperately needed now.

Although disagreements exist in science, the discipline is self-correcting as others attempt to repeat or question research. Sensational claims with enormous implications are subject to intense scrutiny and eventual correction, modification, or rejection. The scientific description of the state of the physical, chemical, and biological components of the planet, agreed upon by scientists in seventy-one countries in 1992, then 184 nations in 2017, attests that scientific information transcends the biases imposed by religion, politics, economics, and laws and thereby guides all of humanity on a path to long-term survival.

A brilliant sun star in July 2022.

Peter A. Victor

ECONOMIST, PROFESSOR EMERITUS AND SENIOR SCHOLAR AT YORK UNIVERSITY

I had the unusual good fortune of meeting David on two very different occasions. The first was in 1967. I was at the frosh event for new students at UBC. I had just arrived from the U.K. for graduate studies in economics and took every opportunity to meet new people. As I wandered among the crowd, I noticed a group hanging on every word spoken by a flamboyantly dressed Asian man, with masses of hair bound by a colourful bandana. He was making the case for a moratorium on nuclear power. Although I didn't speak to him, I did think we had a meeting of minds and was excited by the combination of knowledge and passion for which David was to become famous.

Decades later, I was invited to meet with David, this time for real, to discuss the possibility of joining the board of the David Suzuki Foundation. We met in the grand foyer of the CBC building in Toronto and found a quiet spot to talk. We were enjoying a lively conversation when a man approached holding a young child in his arms. Without introducing himself, he addressed David saying, "Dr. Suzuki, you must save the world for the sake of our children." What struck me was not just the courteous way in which David responded but the enormity of the expectations placed upon him by complete strangers. I went on to serve nine years as a member of the foundation's board and came to appreciate, as so many other have done, what a dedicated and wise person we have in our midst. May he be with us for many, many years to come.

Helena Norberg-Hodge

FOUNDER AND DIRECTOR OF LOCAL FUTURES, AUTHOR, FILMMAKER

In the early 1990s, we released a film, *Ancient Futures,* which pointed to the globalized economy as the driver of our ecological crises. It was a time when most environmental organizations—and, of course, the mainstream media—shied away from any critique of the economy. But David saw the need for environmentalists to focus on an economic shift and had the courage to air *Ancient Futures* on *The Nature of Things*. We have been friends and soulmates ever since.

THE DAVID SUZUKI INSTITUTE is a companion organization to the David Suzuki Foundation, with a focus on promoting and publishing on important environmental issues in partnership with Greystone Books.

We invite you to support the activities of the Institute. For more information, please write to us at info@davidsuzukiinstitute.org.